Emerging Applications of Vacuum-Arc-Produced Plasma, Ion and Electron Beams

NATO Science Series

A Series presenting the results of scientific meetings supported under the NATO Science Programme.

The Series is published by IOS Press, Amsterdam, and Kluwer Academic Publishers in conjunction with the NATO Scientific Affairs Division

Sub-Series

I. Life and Behavioural Sciences	IOS Press
II. Mathematics, Physics and Chemistry	Kluwer Academic Publishers
III. Computer and Systems Science	IOS Press
IV. Earth and Environmental Sciences	Kluwer Academic Publishers
V. Science and Technology Policy	IOS Press

The NATO Science Series continues the series of books published formerly as the NATO ASI Series.

The NATO Science Programme offers support for collaboration in civil science between scientists of countries of the Euro-Atlantic Partnership Council. The types of scientific meeting generally supported are "Advanced Study Institutes" and "Advanced Research Workshops", although other types of meeting are supported from time to time. The NATO Science Series collects together the results of these meetings. The meetings are co-organized bij scientists from NATO countries and scientists from NATO's Partner countries – countries of the CIS and Central and Eastern Europe.

Advanced Study Institutes are high-level tutorial courses offering in-depth study of latest advances in a field.
Advanced Research Workshops are expert meetings aimed at critical assessment of a field, and identification of directions for future action.

As a consequence of the restructuring of the NATO Science Programme in 1999, the NATO Science Series has been re-organised and there are currently Five Sub-series as noted above. Please consult the following web sites for information on previous volumes published in the Series, as well as details of earlier Sub-series.

http://www.nato.int/science
http://www.wkap.nl
http://www.iospress.nl
http://www.wtv-books.de/nato-pco.htm

Series II: Mathematics, Physics and Chemistry – Vol. 88

Emerging Applications of Vacuum-Arc-Produced Plasma, Ion and Electron Beams

edited by

Efim Oks

High Current Electronics Institute,
Tomsk, Russia

and

Ian Brown

Lawrence Berkeley National Laboratory,
Berkeley, U.S.A.

Springer-Science+Business Media, B.V.

Proceedings of the NATO Advanced Workshop on
Emerging Applications of Vacuum-Arc-Produced Plasma, Ion and Electron Beams
Lake Baikal, Russia
24–28 June 2002

ISBN 978-1-4020-1066-8 ISBN 978-94-010-0277-6 (eBook)
DOI 10.1007/978-94-010-0277-6

Printed on acid-free paper

TABLE OF CONTENTS

PREFACE

The NATO-sponsored Advanced Research Workshop (ARW) on "Emerging Applications of Vacuum-Arc-Produced Plasma, Ion and Electron Beams" was held at the Baikal Dunes Resort, Lake Baikal, Russia, on June 24-28, 2002. Participants were from NATO countries Belgium, Czech Republic, Germany, Poland, Turkey and the USA, and from NATO partner countries Bulgaria, Russia, Ukraine and Uzbekistan. The goal of the meeting was to bring together researchers involved in novel applications of plasmas and ion/electron beams formed from vacuum arc discharges, especially in less conventional or emerging scientific areas such as new perspectives on vacuum arc phenomena, generation of high charge state metal ions, heavy ion accelerator injection, multi-layer thin film synthesis, biological applications, generation of high-current high-density electron beams, and more. It was our hope that the meeting would engender new research directions and help to establish new collaborations, prompt new thinking for research and technology applications of vacuum arc science, and in general foster development of the field. The Workshop was a great success, as was clearly felt by all of the attendees. The small number of participants at the meeting tended to encourage a high level of closeness and communication between individuals. The location, a small resort on the western side of Lake Baikal in the vicinity of Irkutsk, was ideal – the isolated location, small and quiet, was excellent and was most conducive to discussion among individuals and small groups quite apart from the formal presentations.

We are greatly indebted to all of those who provided so much help in making the Workshop the success that it was. In particular we'd like to thank Igor Krinberg and Victor Paperny, Irkutsk for their great help with all of the local arrangements. We also thank Gera Yushkov, Scientific Secretary of the ARW, for all of his superb detailed organizing; it was a joy to have Gera as our guide and leader! We are also indebted to Timur Kulevoy for the remarkable job that he did in looking after a contingent of us at Moscow, particularly at the airport in the face of all kinds of problems due to a several-hour-delayed return flight from Irkutsk and consequent missed connections. We would especially like to express our great gratitude to the NATO Science Committee for providing the support that allowed the meeting to take place.

In this Proceedings the papers are listed in the same order, for the most part, as they were presented at the meeting. We are grateful to the authors for the prompt submission of manuscripts, allowing this Proceedings to be published in a timely manner.

Efim Oks
Ian Brown

COHESIVE ENERGY RULE FOR VACUUM ARCS

ANDRÉ ANDERS
Lawrence Berkeley National Laboratory
University of California
1 Cyclotron Road
Berkeley, California 94720-8223, USA
aanders@lbl.gov

Abstract

The Cohesive Energy Rule for vacuum arcs describes an empirical relationship between the cathode material and the arc burning voltage, namely, that the burning voltage depends approximately linearly on the cohesive energy. For Berkeley's vacuum arc ion source system it was quantified as $V = V_0 + A\,E_{CE}$, with $V_0 \approx 14.3$ V and $A = 1.69$ V/(eV/atom). Two arguments are brought forward to identify physical justifications for the empirical rule. First, the self-adjusting burning voltage determines the power input for a given arc current, and therefore the Cohesive Energy Rule connects a material property with the energy conservation law. In order to accomplish the phase transition from the solid to the plasma, energy must be invested, and the cohesive energy represents the energy needed to reach the vapor phase. Only a small fraction of power is directly dissipated in the cathode, and the much larger fraction dissipated in the plasma moves away with the expanding plasma. A possible response of the discharge is to self-adjust the total burning voltage. Through this path, the cohesive energy would affect plasma parameters via the causal chain: cohesive energy – burning voltage – power dissipation – electron temperature – ion charge state and ion kinetic energy. The second reason to justify the rule is that many physical parameters show periodicity as expressed in the Periodic Table of the Elements. Therefore, the periodicity shown by the cohesive energy acts as a proxy for the periodicity exhibited by other quantities, for example, melting and boiling temperatures or ionization energies.

1. Introduction

It is well-known that a cathodic vacuum arc is characterized by plasma production at micrometer-size, non-stationary cathode spots on a globally relatively cold cathode. Research over the last decade has shown that cathode spots have a structure consisting of interacting activity centers called fragments [1], and there is some evidence that fragments may have a substructure that could be called cells [2]. The current density of cathode spot is of order 10^{12} A/m^2, with possibly even high peaks at fragments and cells, and lower values if averaged in time and over the spot-surrounding area. Since the area

E. Oks and I. Brown (eds.),
Emerging Applications of Vacuum-Arc-Produced Plasma, Ion and Electron Beams, 1–14.

of a spot is difficult to precisely define, the current density as a single value may be ill defined; and it may be better to use the concept of a space and time-dependent current density distribution. Even when using the simplified current density approach, we realize that the associated areal power density is of order 10^{13} W/m^2 because the cathode fall is of order 20 Volts. This power density is sufficient to transform cathode material from the solid to the plasma phase in an extremely short time period of 10-100 ns [1]. It is appropriate to call a phase transition of that short duration an explosion. The nanosecond time range constitutes the shortest characteristic time of cathodic arcs. Fluctuations on longer time scales also appear due to the statistical superposition of numerous elementary events. Vacuum arc operation can be understood in the frame of cyclic ignition and decay of emission centers; so-called explosive electron emission [3] makes use of the synergetic combination of high electric field and high local temperature. Taking the non-stationary nature of emission one step further, Mesyats [4] called these characteristic emission events "ectons" – implying the picture of quantum-like power portions that are needed for a microexplosion to function.

As long as the cathode is cold, and thus is unable to provide significant thermionic electron emission current, explosive emission is a necessity for arc operation, regardless of cathode material. However, each material has certain characteristics such as "the ease to burn," or, more precise, the likelihood to spontaneously extinguish, the level of fluctuations in the burning voltage, light emission, ion charge state distribution, ion energies, etc. In this paper, these more-or-less subtle differences between cathode materials are discussed in terms of empirical rules. Among them, the Cohesive Energy Rule stands out because it is based on fundamental considerations that include energy conservation and power distribution.

2. The Cohesive Energy Rule

The Cohesive Energy Rule [5, 6] can be formulated as follows: "The average arc burning voltage of a vacuum arc at a given current is approximately directly proportional to the cohesive energy of the cathode material." The arc burning voltage determines the energy dissipated for a given current, which is determined by the electrical discharge circuit. Because the dissipated energy affects practically all plasma parameters, the Cohesive Energy Rule implies a number of secondary rules. This is non-trivial since the cohesive energy is just one of many physical characteristics of a solid, and it is not obvious why it should be well suited to make conclusions about, or to derive predictions for, the parameters of the expanding vacuum arc plasma.

3. Other Empirical Rules

For decades, researchers have tried, with some success, to identify simple relationships that could help to understand cathode spot physics and to predict plasma parameters. For example, Kesaev [7] and Grakov [8] attempted to correlate the burning voltages with arc current and thermophysical properties of the electrode material. Kesaev suggested that there is correlation between arc voltage and the product of the boiling temperature and the square root of the thermal conductivity. Brown and co-workers [9, 10] found a correlation between the boiling temperature of the cathode material (in Kelvin) and the mean ion charge state of the vacuum arc plasma:

$$\bar{Q} = 1 + 3.8 \times 10^{-4} T_{boil} \quad (1)$$

One of the more recent and thorough works was published by Nemirovskii and Puchkarev [11] who derived a complicated relation between burning voltage, the thermal conductivity and the specific heat of the cathode material.

With the large body of experimental data known today it is clear that plasma parameters such as average ion charge state [9, 12], electron temperature [13, 14], and ion velocity [12, 15] can be correlated to periodic properties of the elements as arranged in a Periodic Table.

4. Voltage Measurements: Experimental Evidence for the Cohesive Energy Rule

Measurements of the burning voltage were performed using the vacuum arc ion source "Mevva V" at Lawrence Berkeley National Laboratory. The source has been described elsewhere [16]. The use of this facility has the advantage that the voltage data can be directly associated with other data measured at the same facility, such as ion charge state distributions [10], electron temperatures [14], and directed ion velocities [15, 17]. Furthermore, the ion source facility allows us to measure up to 18 different cathode materials without interrupting vacuum, and thus a survey of many cathode materials can be conducted in an efficient way. The measurements have been reported in a recent paper [6] and therefore only a short summary is given here.

The "revolver barrel" electrode holder of the ion source has an intrinsic resistance that systematically contributes to the measured voltage data. An additional voltage contribution occurs through the resistivity of the cathode itself. Therefore, the measured raw voltage data do not represent the true burning voltage but include all voltage contributions between the measuring points, i.e. the total voltage drop between anode and cathode connector.

Corrections to the measured data were obtained by using the same setup without plasma, i.e. by measuring the voltage for a shorted cathode-anode gap. These measurements were not done in vacuum but in air, or under inert gas when using reactive cathode materials. The difference between the voltage measured with plasma and without plasma (shorted gap) is approximately equal to the burning voltage. It is not precisely equal since the shorted gap situation includes the small but non-zero contact resistance of the short. The error in the measurements increases with increasing current. The error is less for noble metals (such as Au, Pt) and may be higher for very reactive metals (such Li, Ca, Sr, Ba).

KEY

element number	**22 Ti** (**element symbol**)
cohesive energy (eV/atom)	4.85
average burning voltage (V)	21.3
electron temperature (eV)	3.2
mean ion charge state	2.03
average ion kinetic energy (eV)	58.9

1 H								
3 Li 1.63 23.5 2.0 1.00 19.3	**4 Be** 3.32 - 2.1 - -							
11 Na 1.113 - 1.8 - -	**12 Mg** 1.51 18.8 2.6 1.54 49.4							
19 K 0.934 - 1.7 - -	**20 Ca** 1.84 23.5 2.2 1.93 39.9	**21 Sc** 3.9 19.1 2.4 1.79 49.6	**22 Ti** 4.85 21.3 3.2 2.03 58.9	**23 V** 5.51 22.5 3.4 2.14 70.2	**24 Cr** 4.1 22.9 3.4 2.09 71.6	**25 Mn** 2.92 22.0 2.6 1.53 -	**26 Fe** 4.28 22.7 3.4 1.82 45.9	**27 Co** 4.39 22.8 3.0 1.73 44.4
37 Rb 0.852 - - - -	**38 Sr** 1.72 18.0 2.5 1.98 60.5	**39 Y** 4.37 18.1 2.4 2.28 80.3	**40 Zr** 6.25 23.4 3.7 2.58 112	**41 Nb** 7.57 27.0 4.0 3.00 128	**42 Mo** 6.82 29.53 4.5 3.06 149	**43 Tc** 6.85 - - - -	**44 Ru** 6.74 23.8 - 2.9 139	**45 Rh** 5.75 24.8 4.5 3.0 142
55 Cs 0.804 - - - -	**56 Ba** 1.90 18.3 2.3 2.00 44.6	**57 La*** 4.47 17.2 1.4 2.22 34.6	**72 Hf** 6.44 24.3 3.6 2.89 97.5	**73 Ta** 8.10 28.7 3.7 2.93 136	**74 W** 8.99 31.9 4.3 3.07 117	**75 Re** 8.03 - - - -	**76 Os** 8.17 - - - -	**77 Ir** 6.94 24.5 4.2 2.66 113
87 Fr - - - -	**88 Ra** 1.66 - - -	**89 Ac**** 4.25 - - -						

* Lanthanides	**58 Ce** 4.32 17.9 1.7 2.11 45.5	**59 Pr** 3.70 20.0 2.5 2.25 51.5	**60 Nd** 3.40 19.7 1.6 2.17 49.7	**61 Pm** - - - - -	**62 Sm** 2.14 14.6 2.2 2.13 51.8	**63 Eu** 1.86 21.3 2.0 - -
** Actinides	**90 Th** 6.20 23.3 2.4 2.88 118	**91 Pa** - - - - -	**92 U** 5.55 23.5 2.4 3.18 160	**93 Np** 4.73 - - - -	**94 Pu** 3.60 - - - -	**95 Am** 2.73 - - - -

								2 He
			5 B 5.81 - 2.1 - -	**6 C** 7.37 29.6 2.0 1.00 18.7	**7 N**	**8 O**	**9 F**	**10 Ne**
			13 Al 3.39 23.6 3.1 1.73 33.1	**14 Si** 4.63 27.5 2.0 1.39 34.5	**15 P** 3.43	**16 S** 2.85	**17 Cl**	**18 Ar**
28 Ni 4.44 20.5 3.0 1.76 40.6	**29 Cu** 3.49 23.4 3.5 2.06 57.4	**30 Zn** 1.35 15.5 2.0 1.20 35.7	**31 Ga** 2.81 - 2.0 - -	**32 Ge** 3.85 17.5 2.0 1.40 46.2	**33 As** 2.96	**34 Se** 2.46	**35 Br**	**36 Kr**
46 Pd 3.89 21.3 3.5 1.88 80.1	**47 Ag** 2.95 23.6 4.0 2.14 68.7	**48 Cd** 1.16 16.0 2.1 1.32 26.6	**49 In** 2.52 17.5 2.1 1.34 21.6	**50 Sn** 3.13 17.5 2.1 1.53 29.5	**51 Sb** 3.14 17.5 1.4 1.00 -	**53 Te** 2.19	**53 I**	**54 Xe**
78 Pt 5.84 22.5 4.0 2.08 67.2	**79 Au** 3.81 19.7 4.0 1.97 49.0	**80 Hg** 0.67 - 2.3 - -	**81 Tl** 1.88 - 2.3 - -	**82 Pb** 2.03 15.5 2.0 1.64 35.8	**83 Bi** 2.18 15.6 1.8 1.17 23.9	**84 Po** 1.50	**85 At**	**86 Rn**

64 Gd	**65 Tb**	**66 Dy**	**67 Ho**	**68 Er**	**69 Tm**	**70 Yb**	**71 Lu**
4.14	4.05	3.04	3.14	3.29	2.42	1.60	4.43
21.6	18.1	19.8	20.0	19.0	21.7	14.4	-
1.7	2.1	2.4	2.4	2.0	2.6	2.2	-
2.20	2.2	2.30	2.30	2.36	1.96	2.03	-
54.1	58.1	59.4	64.1	69.3	-	-	-

Table I. Cohesive Energy [18] in (eV/atom), average arc burning voltage in V [6] at 300 A arc current, mean ion charge state for ~150 μs after arc triggering [19], and average kinetic ion energy [17] in eV for most conductive elements of the Periodic Table.

The voltage measurements between anode and cathode connector were performed using calibrated voltage divider probes (Tektronix P6139A) and a digital oscilloscope (Tektronix TDS 744). Voltage calibration was verified against a calibrated voltage source. The time response was checked using a 50 MHz pulse generator (Tektronix TM 503). The standard electrical circuit of the arc supply of the ion source was used, i.e., an 8 stage pulse forming network providing arc pulses of 250 μs duration, constant (but adjustable) current amplitude of up to 600 A, with typically 2 pulses per second. The burning voltage had an initial peak due to ignition processes and due to the inductive component of the circuit including electrode and plasma inductance. After about 150 μs, the burning voltage for a given cathode material and arc current shows a flat, almost constant value. It is important to note that the voltage value is not the cathode fall and not the minimum burning voltage often reported in literature. It is rather an average voltage where voltage spikes and fluctuations contribute to the average. Therefore, despite corrections derived from the short-circuit measurements, the data reported in Table I are somewhat higher than some literature data.

Table I also shows the cohesive energy expressed in eV/atom. The cohesive energy is the energy needed to form a free, electrically neutral atom from the solid. It is often given in kJ/mol or kcal/mol or J/g, but it can be expressed in eV/atom, representing the average binding energy of the atom in the solid. Data for the cohesive energy (at 0 Kelvin) can be found in many textbooks on solid state physics (e.g. [18]). Figure 1 shows the correlation between arc voltage and cohesive energy.

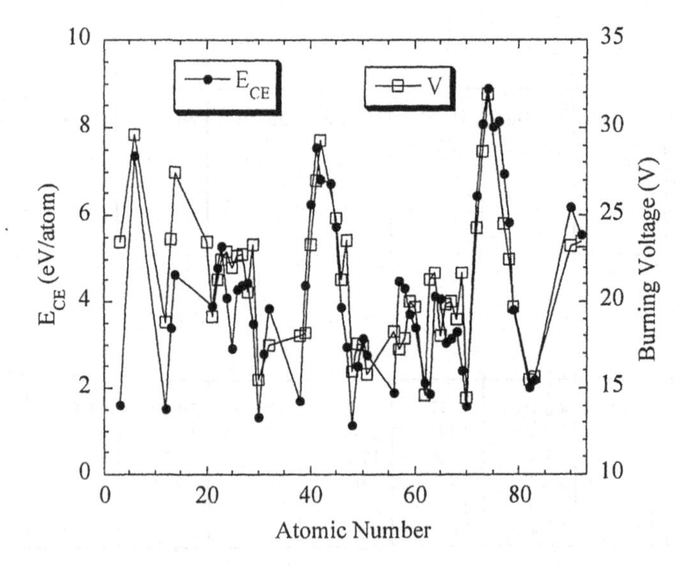

Figure 1. Cohesive energy, E_{CE}, and arc burning voltage, V, for cathode materials of atomic number Z. E_{CE} was taken from ref.[18], and the burning voltage was measured at 300 A as shown in Table I.

5. What Is the Physical Justification for the Cohesive Energy Rule?

Vacuum arcs are different from gaseous arcs in that the arc action needs to provide the material for current transport between electrodes. The transition from the cathode's solid phase to the plasma phase is an essential part of the process. It requires energy, which is supplied via the power dissipated by the arc,

$$P_{arc} = V\, I_{arc} \,. \tag{2}$$

The energy needed for the phase transitions is only a fraction of the total energy balance. The total balance can be written as

$$I_{arc}\, V\, \tau = E_{phon} + E_{CE} + E_{ionization} + E_{kin,i} + E_{ee} + E_{th,e} + E_{MP} + E_{rad} \,, \tag{3}$$

where τ is a time interval over which fluctuations are averaged, E_{phon} is phonon energy (heat) transferred to the cathode material, E_{CE} is the cohesive energy needed to transfer the cathode material from the solid phase to the vapor phase, $E_{ionization}$ is the energy needed to ionize the vaporized cathode material, $E_{kin,i}$ is the kinetic energy given to ions due the pressure gradient and other acceleration mechanisms, E_{ee} is the energy needed to emit electrons from the solid to the plasma (latent work function), $E_{th,e}$ is the thermal energy (enthalpy) of electrons in the plasma, E_{MP} is the energy invested in melting, heating, and acceleration of macroparticles, and E_{rad} is the energy emitted by radiation. The various terms of Equ. (3) contribute very differently to the balance, and some of the terms such as energy invested in macroparticles and radiation are small. The input energy is mostly transferred to heat the cathode, to emit and heat electrons, and to produce and accelerate ions. The cohesive energy E_{CE} is relatively small and may be neglected if one wants to calculate the more prominent energy contributions. However, the correlation between burning voltage and cohesive energy suggests to have a closer look on the physical situation.

There are two arguments to hold up the Cohesive Energy Rule despite the relatively small fraction of energy needed for the phase transition.

The first argument is based on the spatial distribution of the energy input. Figure 2 schematically shows that most of the dissipated energy is concentrated near the cathode surface and associated with the cathode fall. Interestingly, driven by the extreme pressure gradient, both electrons and ions are accelerated away from the cathode surface, carrying away the energy invested in them. Therefore, a very large fraction of input energy is not available to accomplish the phase transition. Materials with large cohesive energy require more energy for the transition from the solid to the vapor phase. The discharge system can provide this greater energy to the solid by enhancing the overall burning voltage. Of course, when this happens, the fractions of energy going to other energy forms, as described by Equ. (3), increase as well. If this is true, solids with greater cohesive energy have higher burning voltage and more energy available for ionization and acceleration of ions, in agreement with measurements.

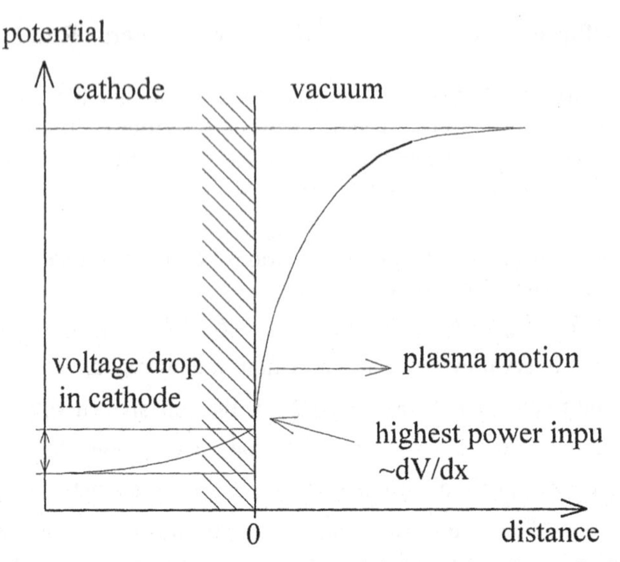

Figure 2. Schematic illustration of the potential drop near the cathode surface. The dissipated power is proportional to the gradient of the potential. The dense plasma, heated most by the power input, is rapidly accelerated away from the cathode, thereby removing most of the input energy.

The second argument for the validity of the Cohesive Energy Rule and derived rules is based on the periodicity of many material properties. Periodicity refers here to properties that are due to the electronic shell structure and that can be grouped according to the Periodic Table of the Elements. For example, the melting temperature and boiling temperature show the same periodicity as the cohesive energy, which of course is not coincidental but related to the bond strength (interaction energy). The periodicity is the physical reason that a relation identified for the melting or boiling temperature, for example, can be formulated as a relation to the cohesive energy. One may argue that that a relation identified for the boiling temperature, for example, is equivalent to a relation to the cohesive energy. That is not entirely correct because relations based on the cohesive energy can be associated with the thermodynamic law of energy conservation. Energy is physical quantity for which we can formulate a balance equation, but temperature is not.

6. Quantification of the Cohesive Energy Rule

In a zero-order approximation one can state that the vacuum arc burning voltage is about 20 V, with somewhat lower values for cathode materials of smaller cohesive energy, and higher values for materials of greater cohesive energy. Quantifying the Cohesive Energy Rule in a first-order approximation gives

$$V = V_0 + A\, E_{CE} \tag{4}$$

where, for the specific experiments with 300-A arcs in a vacuum arc ion source, the values $V_0 = 14.3$ V and $A = 1.69$ V/(eV/atom) were found [6]. For other arc configurations, and considering threshold currents rather than 300 A arcs, the constant will be smaller, perhaps as low as $V_0 \approx 8$ V. Figure 3 shows how well the simple approximation (4) represents the Cohesive Energy Rule.

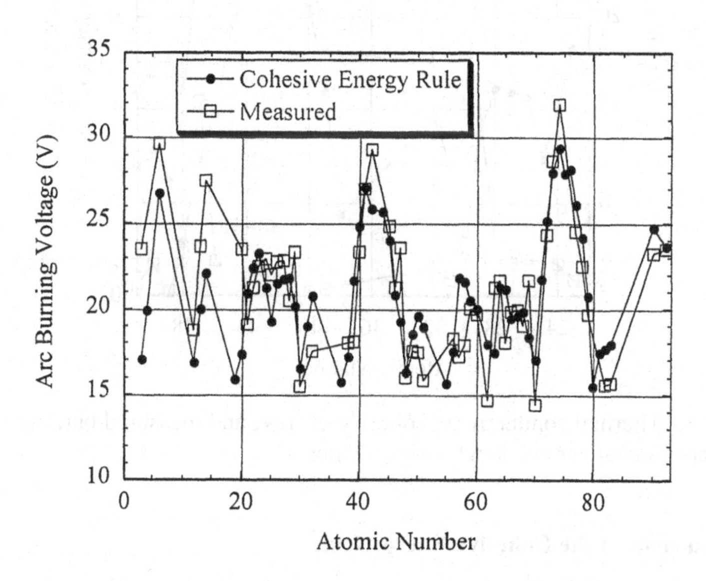

Figure 3. Arc burning voltages for elements of various atomic numbers; one data set was measured (Table I) and the other was calculated using Equ.(4). The lines serve only for the purpose of guiding the eye. Note that a few more elements have been calculated than measured.

One may construct higher-order approximations taking into account properties that affect the energy balance. For instance,

$$V = V_0 + W_1 \, A \, E_{CE} + W_2 \, B \, \lambda \tag{5}$$

adds a term for the thermal conductivity; W_1 and W_2 are weight factors one would normalize by $W_1 + W_2 = 1$, and the factor B could be defined as $B = \left(\lambda - \bar{\lambda}\right)/\bar{\lambda}$ where $\bar{\lambda} \approx 74 \text{ W K}^{-1} \text{ m}^{-1}$ is an average thermal conductivity used for normalization. The small voltage deviations for copper and silver (Figure 4) can be readily fit this way but it turns out that the cohesive energy term is indeed very dominant ($W_1 = 0.997$ and $W_2 = 0.003$). A possible physical interpretation is that the thermal conductivity is not very important for the energy balance due to the explosive nature of the material's phase transition: Before significant amounts of heat are conducted, the spot already "lives" at a different place.

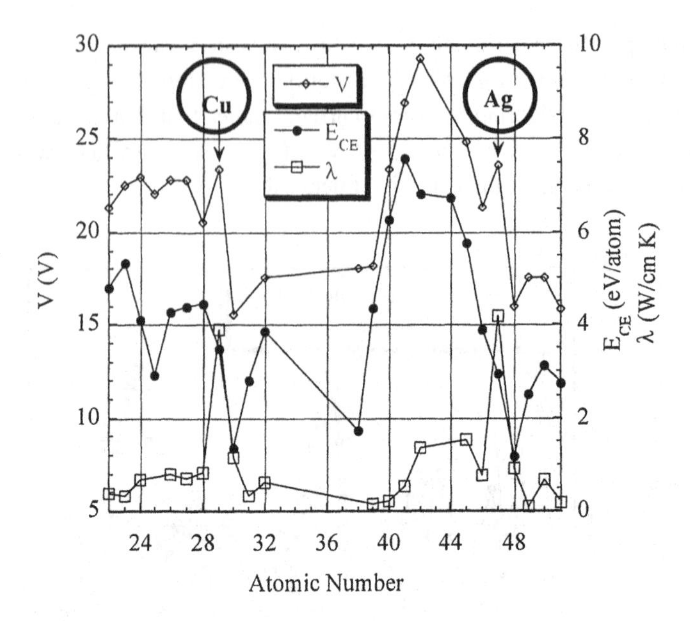

Figure 4. Thermal conductivity, cohesive energy, and measured burning voltage for copper and silver and neighboring elements.

7. Consequences of the Cohesive Energy Rule

Systematic voltage measurements and energy balance considerations allowed us to identify a linear relation between the cohesive energy and the arc burning voltage, the Cohesive Energy Rule. In section 5, two arguments were brought forward to put the empirical rule on a physical basis.

The importance of the first argument, which was based on specifics of the spatial energy distribution in relation to the energy requirement for the solid-plasma phase transition, is that a bewildering variety of cathode processes and parameters has been reduced to a simple number that is a fundamental material characteristics. This simple number is the cohesive energy, a material property that exists unconditionally without any feedback from plasma processes but is determined by the electronic shell structure of atoms. Because a physical justification could be identified based on energy balance considerations and particle and energy fluxes due to pressure gradients, the Cohesive Energy Rule appears to have a deeper physical meaning than other empirical rules mentioned in section 3. However, one should realize that of course complicated processes occur indeed. The Cohesive Energy Rule is a useful rule, not a physical law. Better justification could imply that other empirical rules could be considered as derived from, or secondary to the Cohesive Energy Rule.

The second argument in section 5 was based on the periodicity of the many atomic and solid and liquid state parameters due to the electronic shell structure of atoms. It was argued that the periodicity in one parameter, like the cohesive energy, is a proxy for the periodicity of other parameters. Figure 5 shows, for example, the strong correlation between cohesive energy and boiling point of the cathode. A consequence of the similar periodicity is that all empirical rules of the vacuum arc reflect the same physics, namely, that physical processes at the cathode spot are affected by a number of parameters which happen to show similar periodicity. Still, as discussed above, the Cohesive Energy Rule should be preferred since it deals with energy (a quantity with a conservation law) rather than temperature or other properties.

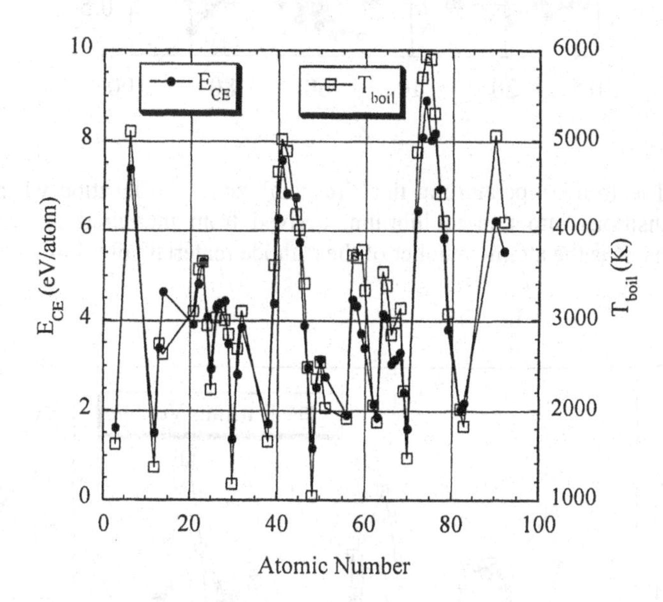

Figure 5. Cohesive energy, E_{CE}, and the boiling temperature of cathode materials of atomic number Z.

From the Cohesive Energy Rule one may consider the following causal sequence of influences:

cohesive energy ⇨ burning voltage ⇨ power dissipation ⇨ electron temperature ⇨ ion charge state and ion kinetic energy

This sequence is well supported by the experimental data of Table I. For example, Figures 6 and 7 illustrate the correlation between charge states, electron temperature, and burning voltage.

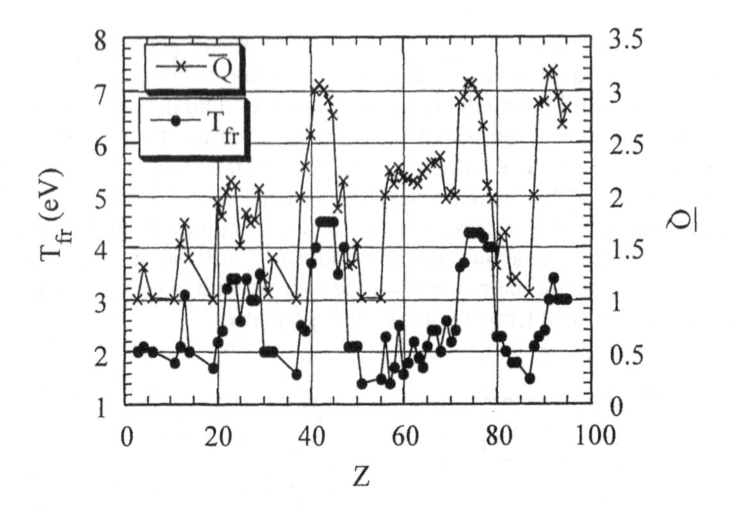

Figure 6. Electron temperature at the "freezing" zone, i.e. location where the spot plasma transitions into non-equilibrium, derived from measured ion charge state distributions; Z is the atomic number of the cathode material, after [14].

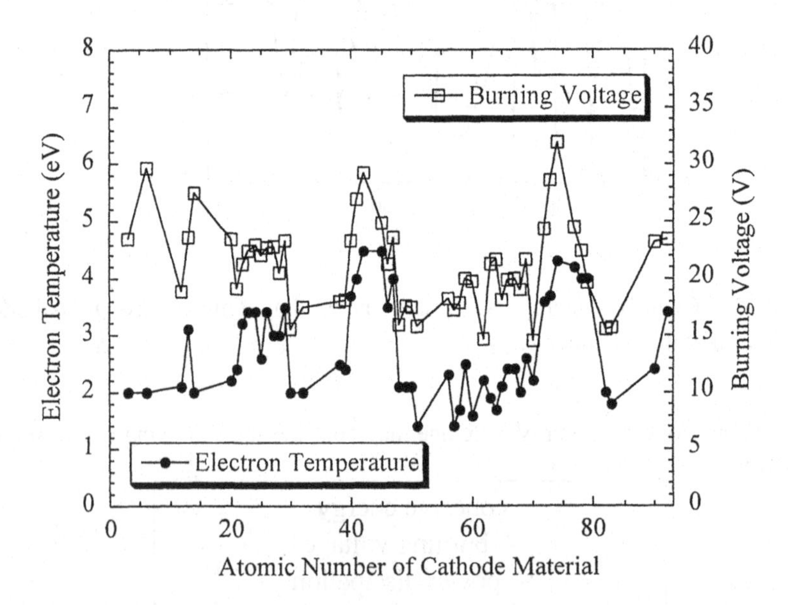

Figure 7. Burning voltage and plasma electron temperature (in the transition zone to non-equilibrium) as a function of the cathode material. A correlation is evident (lines are only drawn to guide the eye).

Other features of vacuum arc discharges can also be related to the Cohesive Energy Rule. For example, the ease of phase transition can be correlated to arc stability and the likelihood of arc chopping.

Kesaev [20] found a correlation between arc stability and melting point of the cathode material. Arc stability was measured by the average arc duration for a given (relatively low) arc current. Kesaev further noticed that the melting point is not sufficient but the atomic weight of the material should also be an important parameter. Farrall [21] questions this argument by pointing out that the cathode's vapor pressure rather than the atomic weight is the ordering parameter. As an example, Farrall states that copper and beryllium have very different atomic weight but show nearly identical arc stability and vapor pressure. According to Farrall [21], the order of decreasing arc stability is Zn, Bi, Ag, Cu, Mo, W for these selected materials and arc currents ~10 A, with the vapor pressure having the same order. For the three metals Cd, Be, and Mo, the chopping current increases with decreasing vapor pressure.

All of the above findings are in qualitative agreement with predictions by the Cohesive Energy Rule. The smaller the cohesive energy the easier are the phase transitions and thus the lower is the chopping current and the more stable is the arc discharge. However, a word of caution is prudent: arc stability and chopping current depend also on the surface state of the cathode, and therefore is it difficult to compare data without the exact knowledge of the surface conditions.

8. Conclusions

The Cohesive Energy Rule gives simple, empirical guidance on a number of arc plasma parameters despite the complex nature of arc spot processes. However, because details of these processes are not considered, the Cohesive Energy Rule is neither a physical law nor can it replace detailed modeling. Although empirical, the Cohesive Energy Rule has its roots in energy conservation and distribution, coupled with the periodic structure of electronic shells of atoms. Therefore, the Cohesive Energy Rule should be preferred to other empirical rules that have been found for vacuum arcs.

9. Acknowledgments

Ian G. Brown is gratefully acknowledged for his long-term support and for presenting this work at the NATO workshop. Gera Yu. Yushkov and Banchob Yotsombat have contributed to ion energy and voltage measurements. This work was supported by the U.S. Department of Energy, under Contract No. DE-AC03-76SF00098.

10. References

[1] B. Jüttner, "Cathode spots of electrical arcs (Topical Review)," *J. Phys. D: Appl. Phys.*, vol. 34, pp. R103-R123, 2001.

[2] I. Kleberg, " Dynamics of cathode spots in external magentic field (in German)," in *Naturwissenschaftliche Fakultaet*. Berlin, Germany: Humboldt University, 2001.

[3] G. A. Mesyats and D. I. Proskurovsky, *Pulsed Electrical Discharge in Vacuum*. Berlin: Springer-Verlag, 1989.

[4] G. A. Mesyats, *Cathode Phenomena in a Vacuum Discharge: The Breakdown, the Spark, and the Arc*. Moscow, Russia: Nauka, 2000.

[5] A. Anders, "Energetics of vacuum arc cathode spots," *Appl. Phys. Lett.*, vol. 78, pp. 2837-2839, 2001.

[6] A. Anders, B. Yotsombat, and R. Binder, "Correlation between cathode properties, burning voltage, and plasma parameters of vacuum arcs," *J. Appl. Phys.*, vol. 89, pp. 7764-7771, 2001.

[7] I. G. Kesaev, "Laws governing the cathode drop and the threshold currents in an arc discharge on pure metals," *Sov. Phys. - Techn. Phys.*, vol. 9, pp. 1146-1154, 1965.

[8] V. E. Grakov, "Cathode fall of an arc discharge in a pure metal," *Sov. Phys. - Techn. Phys.*, vol. 12, pp. 286-292, 1967.

[9] I. G. Brown, B. Feinberg, and J. E. Galvin, "Multiply stripped ion generation in the metal vapor vacuum arc," *J. Appl. Phys.*, vol. 63, pp. 4889 - 4898, 1988.

[10] I. G. Brown and X. Godechot, "Vacuum arc ion charge-state distributions," *IEEE Trans. Plasma Sci.*, vol. 19, pp. 713-717, 1991.

[11] A. Z. Nemirovskii and V. F. Puchkarev, "Arc voltage as a function of cathode thermophysical properties," *J. Phys. D: Appl. Phys.*, vol. 25, pp. 798-802, 1992.

[12] I. A. Krinberg and M. P. Lukovnikova, "Application of a vacuum arc model to the determination of cathodic microjet parameters," *J. Phys. D: Appl. Phys.*, vol. 29, pp. 2901-2906, 1996.

[13] I. A. Krinberg and M. P. Lukovnikova, "Estimating cathodic plasma jet parameters from vacuum arc charge state distribution," *J. Phys. D: Appl. Phys.*, vol. 28, pp. 711-715, 1995.

[14] A. Anders, "Ion charge state distributions of vacuum arc plasmas: The origin of species," *Phys. Rev. E*, vol. 55, pp. 969-981, 1997.

[15] G. Y. Yushkov, A. Anders, E. M. Oks, and I. G. Brown, "Ion velocities in vacuum arc plasmas," *J. Appl. Phys.*, vol. 88, pp. 5618-5622, 2000.

[16] I. G. Brown, "Vacuum arc ion sources," *Rev. Sci. Instrum.*, vol. 65, pp. 3061-3081, 1994.

[17] A. Anders and G. Y. Yushkov, "Ion flux from vacuum arc cathode spots in the absence and presence of magnetic fields," *J. Appl. Phys.*, vol. 91, pp. 4824-4832, 2002.

[18] C. Kittel, *Introduction to Solid State Physics*. New York: John Wiley & Sons, 1986.

[19] A. Anders, "A periodic table of ion charge-state distributions observed in the transition region between vacuum sparks and vacuum arcs," *IEEE Trans. Plasma Sci.*, vol. 29, pp. 393-398, 2001.

[20] I. G. Kesaev, "Stability of metallic arcs in vacuum," *Sov. Phys. - Techn. Phys.*, vol. 8, pp. 457-462, 1963.

[21] G. A. Farrall, "Current zero phenomena," in *Vacuum Arcs Theory and Application*, J. M. Lafferty, Ed. New York: Wiley, 1980, pp. 184-227.

PHYSICAL BASIS OF PLASMA PARAMETERS CONTROL IN A VACUUM ARC

I.A. Krinberg
Irkutsk State University
Irkutsk, 20 Gagarin blvd,
664000, Russia

Abstract. Validity of theoretical models of the arc plasma expansion into vacuum is discussed. It is shown that the model of free spherical expansion is in a close agreement with measurements at low currents but check of the theory of magnetically confined expansion is incomplete due to the absence of regular measurements of plasma parameters at moderate arc currents. The models discussed and developed, are used to interpret experimental data on the vacuum arc plasma. Interpretation of a few effects observed is given and prediction of arc plasma behaviour is made.

1. Introduction

In different technical applications it is usually desirable to have high kinetic ion energy and high ion charge states. It is known that the variations in energy and charge states originate from changing macroscopic discharge parameters such as current value, current pulse duration, interelectrode gap length, cathode diameter, discharge geometry and so on. The physical phenomena resulting from discharge parameters variation are discussed.

The physical processes in a vacuum arc may be divided into two groups. The primary ones occur in the cathode spots and the secondary ones take place in the interelectrode gap. The cathode spot processes are spontaneous and chaotic. The feasibility of their control is restricted by the cathode material choice and the electrode surface cleaning. These actions enable to vary the mean ion charge Z and the ion energy $E = mV^2/2$ in a relatively narrow range of Z =1-3 and E =20-200 eV respectively [1-3]. The secondary processes are more regular and one has some possibilities of ion parameters controlling in a range of Z =3-7 and E =1-10 keV [4-6]. At a very short current pulse the values of Z =10-20 and E =0.1-1 MeV may be achieved [6-8].

It may be thought that a change of the electron temperature and density is the main reason of variations in ion characteristics because the electron component of vacuum arc plasma is responsible for the acceleration and ionization of ions. Thus in present

E. Oks and I. Brown (eds.),
Emerging Applications of Vacuum-Arc-Produced Plasma, Ion and Electron Beams, 15–26.

work main attention is given to revealing the changes of the electron temperature and density under variations of discharge parameters. For this purpose the different models of a vacuum arc plasma flow are used. An attempt of prediction of arc plasma behaviour and explanation of effects observed is made. Of course, in some cases the theoretical conclusions may be questionable and the experimental measurements in support of the model assumptions are desirable. Formulation of some experiments is discussed.

2. Single-spot, group-spot and multi-group-spot vacuum arc

As it is known [9-11] the cathode spots of widely varying size are usually observed on the cathode surface. Smallest of them are microspots or emission centres (or "ecton") [2]. The classical arc cathode spot (the "true" spot) which can be distinguished on photographs, is named as a macrospot or single spot. The groupspot consists of about 5-20 macrospots [11]. Diameters of microspot, macrospot and groupspot are equal to $D_m \approx 1\text{-}10\ \mu\text{m}$, $D_s \approx 0.2\text{-}0.4$ mm and $D_g \approx 1\text{-}3\text{mm}$ and current per spot lies in the range of 1-10 A, 20-200 A and 200-500 A respectively [2, 9-11].

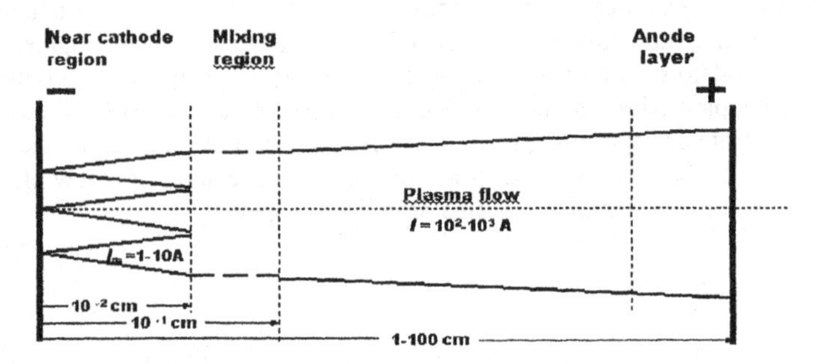

Figure 1. Schematic diagram of plasma expansion in a vacuum arc [6].

In a near cathode region, the vacuum arc plasma exists as a collection of separate micro-jets starting from individual microspots. At the distance r>100 μm the individual micro-jets begin to merge into cathode macro-jet corresponding to a cathode macrospot. When multiple cathode spots are present simultaneously as it is typical at currents exceeding 100 A, some cathode jets may merge into a single plasma jet (a plasma flow) (see Figure 1). The mixing of the cathode macro-jets is a process weakly investigated [12]. Presumably this process occurs at the distances $r_0 \cong 0.1\text{-}1$ mm from cathode surface. According to the structure of plasma production region, the vacuum arc may be labeled as a single-spot, group-spot and multi-group-spot arc. Commonly it is suggested [13] that at low currents (10-100A) the plasma production originates in the vicinity of a single cathode spot but at intermediate currents (0.3-3 kA) several cathode spots exist simultaneously. Correspondingly the low-current vacuum arc is a single-cathode-spot arc and the intermediate current arc is always a multi-cathode-spot arc. But such situation being typical for steady-state vacuum discharge, can not take place in a short pulse arc discharge due to complexity of the cathode spot motion and formation characterized by a wide range of spatial and temporal scales [10,11,14,15]. So, for instance, in a short-pulse discharge of duration $\tau \approx 1 \mu s$ and amplitude current 1-3 kA with Cu-cathode, the authors [16,17] have observed the existence of a single cathode spot, which had no time for its division.

As it has been mentioned above a process of cathode jets merging is weakly investigated. Therefore the area $S_0 = \pi D_0^2 / 4$ of the near-cathode root of united plasma flow is not exactly known value. Naturally it is to put S_0 to be equal to an area occupied by the cathode spots on the electrode surface. For long time operating arc the root area of plasma flow may become equal to the whole cathode surface $S_{cath} = \pi D_{cath}^2 / 4$ (with D_{cath}=1-50 mm) if the arc current is rather high. In other cases the initial plasma flow diameter D_0 is approximately equal to diameter of a single macrospot D_s or a group spot D_g or a size of area $D_{mg} \leq D_{cath}$ occupied by group spots. So in dependence on the cathode spots structure, the initial value of electron density $N_e \propto 1/D_0^2$ may vary in the range of 5 orders of magnitude.

3. Two modes of plasma expansion in a vacuum arc

Two main physical processes control the vacuum arc plasma behaviour: plasma expansion into vacuum ambient and electron migration through the moving ions (i.e. electrical current). First of them results in enhancement of flow diameter accompanying by plasma acceleration and cooling but second one causes plasma heating. So one can expect rapid decreasing of plasma density and increasing of plasma (ions) velocity under plasma motion from the cathode. Indeed it takes place. But a kind of electron temperature variation along plasma flow strongly depends on the rate of plasma expansion.

It is necessary to distinguish two different modes of plasma motion in a vacuum arc: (i) a free expansion into vacuum and (ii) a magnetically confined expansion. In first case plasma pressure essentially exceeds magnetic one. Cross-section of plasma jet (or flow) increases as $S(r) \propto r^2$, electron density decreases as $N_e \propto 1/r^2$ and electron temperature decreases after initial rapid rise [18-24]. Thus ionisation process declines and ion charge state distribution (CSD) freezes. A magnetically confined expansion takes place if the self-magnetic field of discharge current is essential and the magnetic pressure becomes equal to plasma one. In this case plasma flow cross-section increases as $S(r) \propto r$, density decreases as $N_e \propto 1/r$ but electron temperature increases with a distance from cathode [25-27]. So the frequency of recombining collisions decreases but the frequency of ionising collisions can increase due to a temperature rise. Because of plasma cooling the transition from first mode of motion to second one is inevitable. The distance from cathode where such transition takes place, is essentially dependent of arc current value. Therefore first mode is typical for micro-jets and for plasma flow of a low-current arc with interelectrode gap length less than about 10 cm. Second mode is realized at large distances $r \geq 10$ cm for low currents $I \approx 100\text{-}300$ A and rather close to cathode for high currents [6,26]. Additional ionization of ions may occur only under second mode of plasma expansion when enhanced electron temperature T_e becomes higher than primary near-cathode value T_{spot} and a "hot"-zone is formed inside of interelectrode gap [6,28].

4. Modelling of free plasma expansion

A detailed theory has now been derived for the free plasma expansion assumed to be steady state and spherically symmetric in a solid angle $\Omega = 2\pi(1-\cos\alpha)$, where α is the angular size of the plasma jet. The different models [18-24] are in close agreement with each other. All of them demonstrate that electron temperature increases sharp up to the maximum value T_{spot} at the distance $\Delta x << D_m$ from the cathode surface and then decreases fluently. In papers [22,24] the next analytical expressions for arc plasma parameters in a near cathode spot region have been obtained

$$kT_{spot} \approx C_*(ZI/D_m)^{2/5}, \tag{1}$$

$$V_0 \approx 3.5(\gamma Z_0 kT_{spot}/m)^{1/2}. \tag{2}$$

Here $C_* \approx 0.1\,\text{eV(cm/A)}^{2/5}$ holds, V_0 is the ion velocity at distances $(10^2-10^5)D_m$ from cathode, Z_0 is the "frozen" mean ion charge corresponding to the primary ionisation in a region very close to the cathode spot [20,29], m is the ion mass, $\gamma = 5/3$ is the adiabatic index and k is the Boltzmann constant.

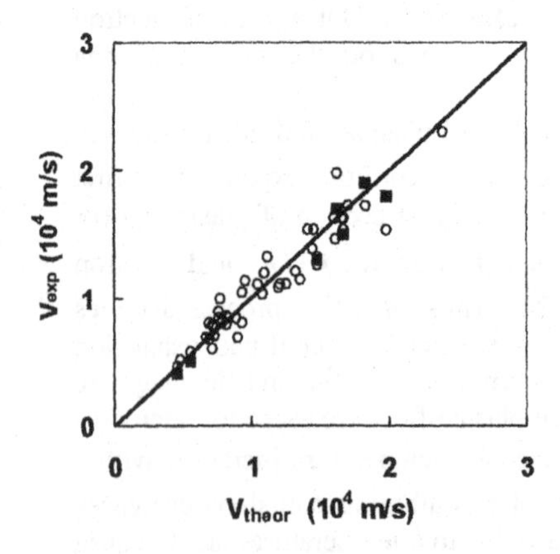

Figure 2. Ion velocity measured in the experiments of [3] (circles) and [31] (squares), versus the ion velocity calculated according Equation (2) for 50 different cathode materials.

Let us check the models developed for validity. Using typical size $D_m \approx 1-3\,\mu\text{m}$ and current per microspot $I \approx 1\text{-}3$ A [2,11] we have $C_1 = C_*(I/D_m)^{2/5} \approx 2\text{-}6$ eV. Then according to equation (1) we obtain $kT_{spot} \approx C_1 Z^{1/2} \approx$ 2-10 eV. So calculated values of electron temperature are close to measured values kT_e =2-6 eV [29,30].

But more direct check of validity of a model may be made with the help of equation (2), which includes the parameters observed experimentally with a good accuracy. At the figure 2 the ion velocities $V_{\exp}$ measured in two series of experiments [3,31], are compared with the velocities $V_{\text{theor}} = V_0$ calculated from equation (2) with values of Z_0 from [1,29] and values of T_{spot} from [29]. High degree of agreement confirms the reality of a model based on assumption of spherical expansion. So other predictions of this model are expected to be correct.

5. Magnetically confined plasma expansion

Self-magnetic field influence on plasma flow parameters has been studied under different assumption in a lot of papers, a short survey of which is given in [6]. On a base of numeric modelling it has been found [25-27,32,33] that magnetic constriction leads to the plasma compression in the near axis region and to the electron temperature enhancement at arc currents $I \geq 300$ A. But results of numeric calculations do not cover a wide range of vacuum arc parameters. Experimental data are also deficient and sporadic. So checking of validity of the models is incomplete. Below an attempt is made to compare the different plasma flow parameters measured at arc current $I \geq 1$ kA, with the calculation results obtained with help of numeric model [6,26] and analytical model developed below.

It is possible to developed simple model considering magnetically confined plasma flow as uniform over its transverse cross section and taking into account variation of the plasma parameters along flow only. For determination of flow cross section S we may suppose that the magnetic force compensates the pressure gradient force on the boundary surface of plasma flow and Bennett criteria is appropriate

$$PS = \mu_0 I^2 / 8\pi, \tag{3}$$

where μ_0 is the magnetic permeability of free space, P is the plasma pressure.

In a steady-state case we have the laws of conservation of charge and mass

$$I = eN_e(V_e - V)S = const, \; G = mN_+ VS = const, \tag{4}$$

equation of ion motion

$$mN_+ V \frac{dV}{dr} = -\frac{dP}{dr}, \tag{5}$$

and equation of electron heat balance

$$\frac{3}{2} P_e V_e \frac{d}{dr} \ln\left(\frac{P_e}{N_e^{\gamma}}\right) = \frac{j^2}{\sigma}. \tag{6}$$

Here r is the distance from cathode, V and N_+ are the velocity and number density of ions; e, P_e and V_e are the charge, pressure and velocity of electrons, $j = I / S$ is the current density. The plasma conductivity is determined by equation $\sigma = k_\sigma T_e^{3/2} / Z$

where $k_\sigma = 3\cdot10^{13}$ eV$^{-3/2}$s^{-1} holds for typical arc plasma parameters. In a case of vacuum arc the electron temperature is more than ion one and the ion pressure can be neglected in comparison with the electron pressure so one has $P \approx P_e = kT_eN_e$. Using the quasineutrality condition $N_e = ZN_+$ and expressing the ion mass flax G in terms of a non-dimensionless erosion coefficient $\delta = eG/mI$ one can find from (4) the production $N_eS = Z\delta I/Ve$. Substituting this value in equation (3) one obtains the expression for electron temperature [34,35]

$$kT_e = C_T(V/Z\delta)I\,, \tag{7}$$

where $C_T = e\mu_0/8\pi =$ $8\cdot10^{-27}$ J·s/A·m=$5\cdot10^{-8}$ eV·s/A·m. At currents $I \leq 500$ A the velocity and the mean charge of ions in a plasma flow are almost the same as in a near cathode region i.e. $V \approx V_0$ and $Z \approx Z_0$. Then those values slightly increase with current rise up to 5 kA but the ratio V/Z remains rather close to V_0/Z_0. So for typical values V_0 =(0.5-2)$\cdot10^4$ m/s (see Fig.1) and $Z_0\delta = \eta =$ 0.07-0.11 [2] we have estimation $kT_e = C_2I$ with $C_2 \approx$ 0.005-0.01 eV/A [34,35].

From equations (3)-(5) it is easy to obtain the next solution

$$V = V_0 + aI\ln(S/S_1) \tag{8}$$

where $a = \mu_0/(8\pi\gamma_i)$, $\gamma_i = G/I = const$ is the erosion coefficient [2]. Solving equations (5),(6) one also obtains

$$T_e^{5/2}S = T_{e1}^{5/2}S_1 + \frac{eI}{k_\sigma}\int_{r_1}^{r}\frac{Zdr}{1+Z\delta}\,. \tag{9}$$

In equations (8),(9) all values labelled by index "1" are related to the point $r = r_1$ beginning of which the magnetic equilibrium approaches. Practically it is possible to put $r_1 \approx r_0$ and $S_1 \approx S_0$.

If the change of electron temperature and mean ion charge in some range of distances is not so large then from equations (7) and (9) one obtains

$$S(r) \approx S_1 + R_S(r - r_1) \cong R_S r\,, \tag{10}$$

$$R_S = eI/T_e\sigma(T_e) = C_R ZI/(kT_e)^{5/2}\,, \tag{11}$$

where C_R = 0.03 cm·eV$^{5/2}$/A. As can be seen from equation (10) the magnetically confined plasma flow has a paraboloid form, that agrees with numeric calculations [25-27].

Figure 3 demonstrates the variation of electron temperature versus distance $X = r - r_0$ where $r_0 \approx$ 0.1-1 mm is the location of mixing region (see Fig.1). One can see the dependence of temperature on discharge current and root size of plasma flow. Agreement with spectroscopic estimation of electron temperature[13,36] is acceptable taking into account uncertainty in a value of flow cross section.

Figure 4 shows the calculated profile of plasma flow for a single-spot vacuum arc compared with measurements in the pulse vacuum discharge [16,17]. Authors have observed formation of a comprehensively narrow glow cylindrical channel of radius $R \approx$ 1-1.5 mm for gap length L =4 mm and $R \approx$ 3-4 mm for L =12 mm throughout the

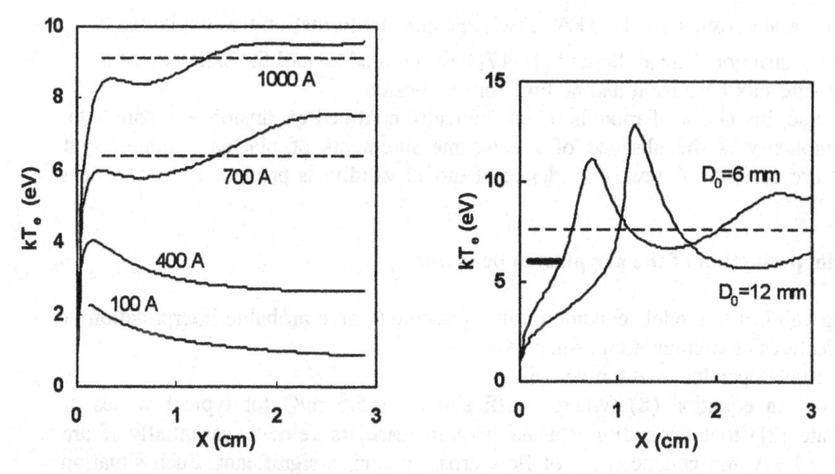

Figure 3. **Distribution of the electron temperature along the axis of plasma flow.**
Narrow solid line – numeric calculation; dashed line -- calculation according equation (7) for equilibrium of plasma and magnetic pressures.
Left side: **Titanium cathode, group-spot arc with $D_0 = D_g$ =2 mm.** *Right side*: **Aluminium cathode, arc current 1 kA, multi-group-spot arc with $D_0 = D_{cath}$ and $D_0 = D_{cath}/2$. Wide solid line – estimation of electron temperature from spectroscopic measurements at I = 1.2 kA and D_{cath} =12 mm [13,36].**

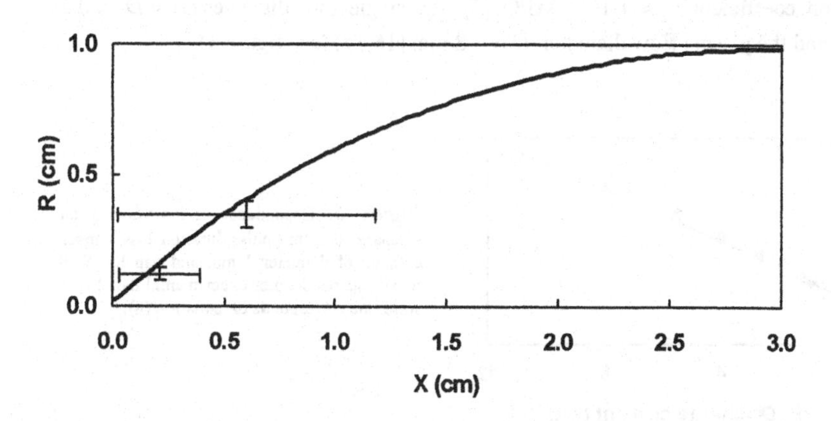

Figure 4. **Profile of the plasma flow for a single-spot vacuum arc with copper cathode. Solid line - calculation results for $D_0 = D_s = 0.3$ mm; horizontal lines with error-bar - measurements [16,17].**

pulse $\tau \approx 1\,\mu s$ at currents $I \approx 1-3$ kA. Discharge channel formation was explained by magnetic constriction. The authors of [16,17] have pointed toward the existence of a single cathode spot because it had no time for its division.

Of course this check of models of magnetically confined expansion is incomplete. Main complexity is the absence of regular measurements of plasma parameters at moderate arc currents. A few of evidence of model validity is presented too in a next section.

6. Model prediction of the arc plasma behavior

Assuming validity of models considered it is possible to give probable interpretation of the next features of vacuum arc plasma behavior.

a) Current dependence of ion velocity

It is seen from equation (8) (where coefficient $a \approx 0.5\text{-}5$ m/Q for typical values of erosion rate [2]) that expanding plasma may increase its velocity essentially if arc current $I \geq 1$ kA and enhancement of flow cross section is significant. Such situation can be realised in a short-pulse discharge when the vacuum arc has possibility to keep a single-spot type. Indeed the velocity increasing was discovered in pulse discharge of duration $\tau \approx 1\mu s$ and amplitude 1-11 kA [37]. Figure 5 shows the existence of linear current dependence of ion velocity $V = V_0 + kI$ with $V_0 \approx 2 \cdot 10^4$ m/s and $k \approx 5$ m/s·A. This agrees well with equation (8). Coefficients obtained are close to the velocity $V_0 \approx 1.3 \cdot 10^4$ m/s, measured in a low-current arc [3], and to the value $k = a\ln(S/S_1) \approx 2a\ln(D/D_0) \approx 7$m/s·A. Last one was calculated for Cu-cathode using the erosion coefficient $\gamma_i \approx 4 \cdot 10^{-8}$ kG/Q [2], the single-spot diameter $D_0 \approx D_s \approx 0.3$ mm [11] and the plasma flow diameter $D \approx 6\text{-}8$ mm [16,17] (see figure 4).

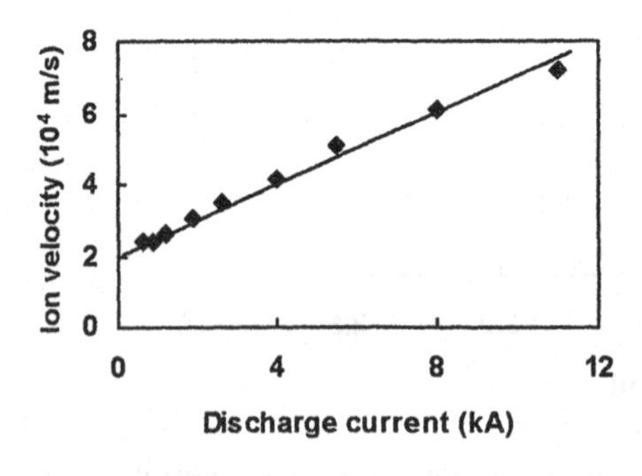

Figure 5. **Dependence of ion velocity on discharge current (pulse duration 1 μs; copper cathode of diameter 1 mm and gap length 9 mm): squares denotes experimental data [37] ; solid line corresponds to equation (8).**

b) Current dependence of ion charge state distribution

As it is seen from equation (7) and figure 3 the electron temperature is proportional to discharge current at $I \geq 300$ A. As a result one may expect increasing of ion charge state with current rise due to the additional ionization by electron impact inside of an interelectrode gap [34-36]. The rise of mean ion charge Z becomes essential if the electron temperature T_e in the region of additional ionisation exceeds the temperature $T_{spot} \approx$ 2-4 eV [29] in the near cathode region. According to equation (7) it takes place for currents $I \geq 500$ A. Indeed a noticeable rise of the mean ion charge state has been observed at the discharge current $I \approx 500-1000$ [4]. Current dependence of the mean ion charge is also illustrated by figure 6 where the results of numerical calculation are given for single-spot, group spot and multi-group spot arcs with titanium cathode.

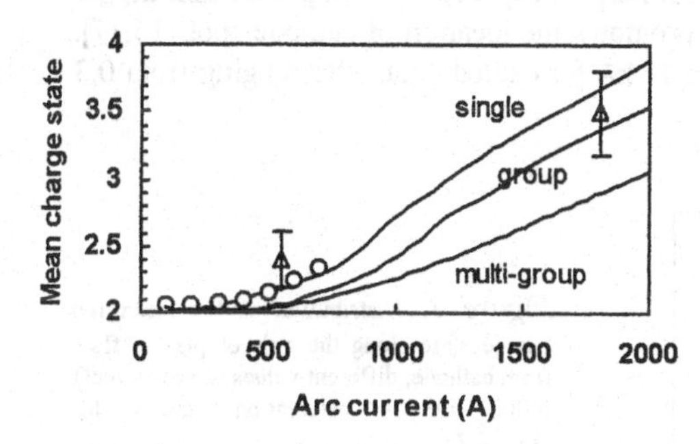

Figure 6. Dependence of mean ion charge state on arc current (titanium cathode). Solid line-calculation results: single-spot arc with $D_0 = D_s$ = 0.4 mm, group-spot arc with $D_0 = D_g$ = 2 mm, multi-group-spot arc with $D_0 = D_{ms} = D_{cath}$ =6 mm. Circles denotes experimental data for steady-state arc [4] and triangles corresponds to measurements for pulse discharge [5].

c) Effect of short current pulse or superimposed pulse

It has been found [5,38] that in a short-pulse discharge ($\tau_{pulse} \approx$ 6-60 μs) the mean ion charge is enhanced by a factor of 1.5-2 compared to a steady-state arc at same discharge current. Similar results have been obtained at a relatively low current (600-800 A) by a "current spike" method when the short current pulse ($\tau_{pulse} \approx$ 5-15 μs) is superimposed on the quasi-stationary arc current [31,39,40]. As it was observed the enhancement of CSD is more essential if the spike duration decreases and there is no measurable enhancement if the spike duration is longer than 200 μs. Next interpretation of this effect may be given for discharge current value $I \geq 500$ A which was realized in experiments mentioned above. If a characteristic time $\tau \approx \tau_{pulse}/2$ of the current rise is more than the ionic time of flight $\tau_L \approx L/V \approx 1$ μs so the properties of such pulse discharge plasma are almost the same as those of a steady-state vacuum arc [28,36]. Thus the electron temperature does not depend on pulse duration and is determined by equation (7). But in a case of short current pulse the vacuum arc can keep a single-spot (or group-spot) type since the cathode spots have no time to be divided and to occupy whole cathode surface. Therefore the electron density $N_e \propto 1/S$ and the ionisation rate will be higher than in a multi-group-spot arc. This conclusion is illustrated by the calculation data given in a figure 6 where the tendency of the ion charge increasing due to the flow size decreasing is obvious. It is interesting to note that a considerable shot-to-shot variation of the CSD is observed for the same amplitude of current [5]. Possible reason is an irregular transition of vacuum arc from single-spot to group-spot mode (see figure 6). Presented interpretation of the effect is also supported by finding [39] that the charge state enhancement in the second current spike is not as great as the one observed for the first spike. Indeed the cathode spots have more time for division and displacement from each other during enhanced temporal interval $\Delta t \approx 2\tau_{pulse}$ than in a case of single current pulse.

d) Influence of the cathode diameter and geometry

Decreasing of cathode diameter till value $D \le D_g \approx 1$ mm may originate some increase of the ion charge state due to the same reason as discussed above i.e. due to restriction of initial size of plasma flow. Similar effect may be expected in setup with thin trigger electrode in the center of cathode which promotes the location of cathode spot [16,17]. Measurements of ion CSD at current $I \cong 1$ kA for cathode diameter ranging from 0.3 to 3 mm would be interesting.

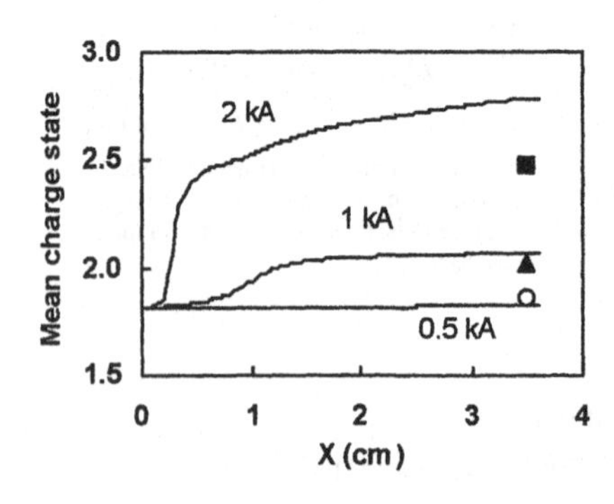

Figure 7. Distribution of the mean ion charge state along the axis of plasma flow (iron cathode, different values of arc current) Solid line - calculation results for $D_0 = D_{ms} = D_{cath} = 6$ mm; circle, triangle and square – results of measurements for same cathode at currents 0.5 , 1 and 2 kA from [4].

e) Influence of the interelectrode gap length

Because the secondary ionization out of a near cathode region determined by time of stay in a "hot" zone, ion charge state will be increasing during ion travelling from cathode to anode. This effect is illustrated by figure 7. It is seen that influence of gap length L ought be more noticeable at lower currents. It would be useful to check this conclusion by means of measurement of ion CSD in the range of $L = 0.3$-3 cm at current $I \cong 1$ kA.

f) Effect of cathode surface cleaning

It is observed [11] that on uniformly eroded cathode, the microspots are mostly separated from one another by distances larger than their diameters. Consequently the merging of microspots is prevented and their mean diameter D_m must be smaller. Thus as it is seen from equation (1), increasing of electron temperature in a near spot region may occur. As a result the enhancement of ion charge state ought take place. Similar effect has been really observed in [41] where the rise of mean ion charge with increasing of arc pulse repetition rate was discovered.

7. Conclusion

Discussion presented shows that models of free spherical plasma expansion are acceptable to describing plasma behavior in the microjets where primary ionization and acceleration of ions occur. Unfortunately it is difficult to control the microspot parameters on cathode surface. There are more possibilities of plasma parameter controlling inside of interelectrode gap where plasma expansion becomes magnetically confined at arc currents more than 0.3 kA. But the validity of model predictions for this case is not checked completely. For example, there are different view-point's on charge state enhancement due to decreasing of pulse duration. What processes lead to increasing of ion charge – primary ionization in a near cathode region [4,5] or secondary ionization inside of an interelectrode gap [6,13,24]? Thus it is desirable to carry out additional measurements of vacuum arc plasma parameters (velocity and charge state of ions, temperature and density of electrons) for discharge gap ranging from 3 to 30 mm with electrode diameter from 0.3 to 3 mm at discharge current of order of 1 kA. Also studying of plasma flow size and form at different parameters of current pulse would be interesting.

Acknowledgments

The author is indebted to E. Zverev for his invaluable help in the numerical solution of a set of the hydrodynamic equations.

References:

1. I. G. Brown and J. E. Galvin, *IEEE Trans. Plasma Sci.* **17**, 679 (1989).
2. G. A. Mesyats, *Cathode Phenomena in a Vacuum Discharge: The breakdown, the Spark and the Arc* Moscow: Nauka (2000).
3. A. Anders and G. Y. Yushkov, *J. Appl. Phys.* **91**, 4824 (2002).
4. E. M. Oks, A. Anders, Brown I G et al., *IEEE Trans. Plasma Sci.* **24**, 1174 (1996).
5. A. Anders, I. G. Brown, M. R. Dickinson, R. A. MacGill. *Rev. Sci. Instrum.*, **67**, 1202 (1996).
6. I. A. Krinberg and V. L. Paperny, *J. Phys. D: Appl. Phys.* **35**, 549 (2002).
7. M. F. Artamonov, V. I. Krasov and V.. L. Paperny, *JETP* **93**, 1216 (2001).
8. E. D. Korop and A. A. Plutto, *Sov. Phys.-Tech. Phys.* **15**, 1986 (1970).
9. J. M. Lafferty, *Vacuum Arcs-Theory and Application* (New York: Wiley, 1980).
10. I. Beilis, B. E. Djakov, B. Juttner and H. Pursch. *J. Phys. D: Appl. Phys.*, **30**, 119 (1997).
11. P. Siemroth, T. Schulke and T. Witke, *IEEE Trans. Plasma Sci.* **25**, 571 (1997).
12. E. Gidalevich, R. L. Boxman, S. Goldsmith, *J. Phys. D: Appl. Phys.*, **31**, 304 (1998).
13. R. L. Boxman and S. Goldsmith , *J. Appl. Phys.* **51**, 3644 (1980).
14. A. Anders, S. Anders, B. Juttner and I.G. Brown, *IEEE Trans. Plasma Sci.*, **21**, 305 (1993).
15. B. Juttner, *IEEE Trans. Plasma Sci.*, **27**, 836 (1999).
16. D. F. Alferov, N. I. Korobova, K. P. Novikova and I. O. Sibiriak, *Proc. XIVth ISDEIV*, Santa Fe, USA 1 542 (1990).
17. D. F. Alferov, N. I. Korobova and I. O. Sibiriak, *Fizika Plasmy* (in Russian) **19**, 399 (1993).
18. B. Ya. Moizhes and V. A. Nemchinskii, *Sov. Phys.-Tech. Phys.* **25**, 43 (1980).
19. C. Wieckert, *Phys. Fluids*, **30**, 1810 (1987).
20. C. Wieckert, *Contrib. Plasma Phys.* **27**, 309 (1987).
21. I. I. Beilis, M. P. Zektser and G. A. Lyubimov, *Sov. Phys.-Tech. Phys.* **33**, 1132 (1988).
22. I. A. Krinberg, M. P. Lukovnikova and V. L. Paperny, *JETP*, **70**, 451 (1990).
23. E. Hantzsche, *IEEE Trans. Plasma Sci.* **20**, 34 (1992).
24. I. A. Krinberg, *Tech. Phys.* **46**, 1371 (2001).

25. I. A. Krinberg and E.A. Zverev, *Tech. Phys. Lett.* **23**, 435 (1997).
26. I. A. Krinberg and E.A. Zverev, *Plasma Phys. Repts.* **25**, 82 (1999).
27. D. L. Shmelev, *Proc. XIXth ISDEIV*, Xi'an, China **1**, 218 (2000).
28. E.A. Zverev and I. A. Krinberg, *Tech. Phys. Lett.* **26**, 288 (2000).
29. A. Anders, *Phys. Rev. E*, **55**, 969 (1997).
30. V. F. Puchkarev, *J. Phys. D: Appl. Phys*. **24**, 685 (1991).
31. A. S. Bugaev, V. I.Gushenets, A. G. Nikolaev, E. M. Oks and G. Y. Yushkov, *IEEE Trans. Plasma Sci.*, **27**, 882 (1999).
32. I. Beilis, M. Keidar, R. L. Boxman and S. Goldsmith, *J. Appl. Phys.* **83**, 709 (1998).
33. I. Beilis and M. Keidar, *Proc. XIXth ISDEIV*, Xi'an, China **1**, 206 (2000).
34. I. A. Krinberg, *Proc. 12th Symp. High Current Electronics*, Tomsk, Russia **1**, 33 (2000).
35. I. A. Krinberg, *Tech. Phys. Lett.* **27**, 45 (2001).
36. S. Goldsmith and R. L. Boxman, *J. Appl. Phys.* **51**, 3649 (1980).
37. S. P. Gorbunov, V. I. Krasov, I. A. Krinberg and V. L. Paperny, *Program XXth ISDEIV*, Tours, France (2002).
38. E. N. Abdullin and G. P. Bazhenov. *Proc. XVIIIth ISDEIV*, Eindhoven, Netherlands, **1**, 207 (1998).
39. G. Y. Yushkov, E. M. Oks, A. Anders and I. G. Brown, *J. Appl. Phys.* **87**, 8345 (2000).
40. A. S.Bugaev, E. M. Oks, G. Y. Yushkov, , A. Anders and I. G. Brown, *Rev. Sci. Instrum.* **71**, 701 (2000).
41. G. Y. Yushkov and A. Anders, *IEEE Trans. Plasma Sci.* **26**, 220 (1998).

SOURCES OF MULTIPLY CHARGED METAL IONS: VACUUM DISCHARGE OR LASER PRODUCED PLASMA?

Victor Paperny
Irkutsk State University,
Irkutsk, Russia

Abstract. An analytical survey is presented of the recent researches aiming to obtain beams of the multiply charged metal ions from vacuum discharges of different types. The main attention is paid to the works, which have been performed the recent two years by the team headed by the author. It is shown that plasma jet of a moderate voltage (< 2.5 kV) vacuum spark generates short- run beams of the cathode matter ions (Cu^{n+}, Ta^{n+}) with the charge states up to +19 for copper and +50 for tantalum. The mean charge states of the ions are: $Z(Cu) = 9.3$, $Z(Ta) = 20$ and current density of these ions attains 30 mA/cm^2 at a distance of 40 cm from the discharge gap. Comparison is performed of these data with the corresponding values for metal ions obtained from a laser produced plasma and it is shown that the vacuum discharge plasma as a source of the multiply charged metal ions is rather more effective than the laser one.

1. Introduction

In recent years, considerable efforts were made for creating sources of multiply charged metal ions, which are of interest for projects of heavy-ion accelerators intended for research purposes, for solving material science problems, and for medical applications. There are two lines of development of these sources. The first one is associated with elaboration of steady-state or 'slow' (with length of about a millisecond) pulsed vacuum arcs. These sources allows to obtain a beams of ions, for instance Pb^{n+}, with the maximum and the mean charge states attaining +7 and +3.4, correspondingly and with the ion current of a few hundreds mA (see, for instance, [1,2]). The second line implies the 'short-run' systems with the length of a few microseconds and less. The most popular type of such sources is a plasma emerging as a result of exposure of a target with a high-power laser pulse, which makes it possible to obtain, for example, Cu^{n+} ions with Z_{max}= +25, Ta^{n+} with Z_{max}= +55 and density of the ion current of about 10 – 20 mA/cm^2 [3,4]. These sources, however, have large dimensions, low pulse repetition rate and high operating cost. Hence, the development of a compact tabletop, low expensive source of highly charged and high-energy metal ions presents an update and urgent problem.

E. Oks and I. Brown (eds.),
Emerging Applications of Vacuum-Arc-Produced Plasma, Ion and Electron Beams, 27–37.

Another promising source of ions is a vacuum spark producing a high energy density released in the plasma column, which is required for generating multiply charged ions and characterized by a relatively low total energy contribution and a simple construction. Indeed, bursts of X rays emitted by hydrogen- or helium-like ions of the cathode material (iron or titanium) from micropinch structures formed in the plasma of a high-current spark were detected even in early experiments (see, for example, [5,6]). Nevertheless, the immediate measurements of ions ejected from plasma of these sparks have found that just the low charge states are present [7]. The experiments have shown also that application a number of methods to increases the ion charge state permitted to produce the peak charge state of ions not exceeding of +7 (Cu^{7+}[8]; Al^{7+}, W^{7+} [9]. These charge states were obtained in high–voltage vacuum sparks with the initial capacitor voltage being of a few tens kilovolts [8] or even of a few hundreds kilovolts [9].

In the present work, we report on the studies of ion beams of the cathode material (Cu^{n+}, Ta^{n+}), generated at the initial stage of burning of a miniature vacuum spark with low values of both the capacitor voltage (up to 2.5 kV) and the stored energy (up to 7 J). The objectives of the studies were follows: (i) to establish if it is possible to obtain beams of the multiply charged ions from a vacuum spark with a relatively low values of the capacitor voltage and the stored energy; (ii) to recognize conditions for the most effective generation of the multiply charged ions in vacuum discharges.

2. Experiment

Experiments were carried out with a low - energy vacuum spark (see Fig.1). The electrode arrangement was mounted in a 100 mm long and 50 mm diameter stainless steel cylindrical chamber. The electrode system consisted of a copper or tantalum cathode of diameter 1 mm and a plane grounded grid-type anode separated by 9 mm from the end face of the cathode and was placed in a chamber with a vacuum not worse than (5-8)· 10^{-6} Torr. The discharge was initiated at the end face of the cathode through the breakdown over the surface of an insulator insert between the cathode and the igniting electrode. The discharge current was sustained by a capacitor ($C = 2\ \mu$F) and was measured by a Rogowki coil directly in the cathode circuit. The total inductance of the discharge circuit did not exceed 40 nH. Before measurements, 'training' during 10^3 'shots' cleaned the cathode surface, after which the variations of discharge parameters in various shots did not exceed 20%.

2.1. ION ENERGY AND CHARGE STATE MEASUREMENTS.

The energy and charge state distributions of ions were measured by the time-of-flight method with the help of an electrostatic analyzer of the 'plane capacitor' type with the energy resolution $\Delta\varepsilon/\varepsilon \approx 2 \times 10^{-2}$ and a time resolution of the registering circuit of about 40 ns. We used a microchannel plate as a detector of ions. The gain of the plate was adjusted for different ion charge states and impact energies using the data obtained in [10]. The analyzer was placed behind the grid anode and the drift gap so that the ions emitted by the cathode flame and moving along the discharge axis towards the anode were registered. The time-of-flight method makes it possible to determine the value of μ/Z (μ is the atomic weight and eZ is the charge of an ion) for each species of ions with an energy specified by the bias voltage at the analyzer plate from the delay of the corresponding signal of the detector.

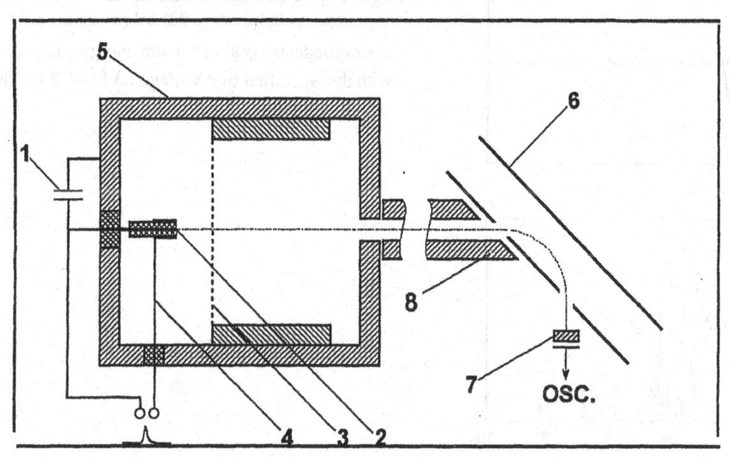

Figure1. The scheme of experiment: 1, capacitor; 2, cathode; 3, anode; 4, igniter, 5, vacuum vessel; 6, electrostatic analyzer of ion energies; 7, ion current detector; 8, drift tube.

Under our experimental conditions and for a gap length of 60 cm, the limiting resolution of the method $a = Z/\Delta Z$ for Cu^{n+} ions was $a(E) = 20$; the corresponding value for Ta^{n+} ions was $a(E) = 30$ in view of their larger mass and, hence, lower velocity. Figure 2(a) shows a typical oscillogram of the discharge current in experiments with a copper cathode; the corresponding signal from the analyzer for a fixed value of E/Z is shown in Fig. 2(b). The signal was processed using the following procedure. Assuming that the last peak in Fig. (b) corresponds to Cu^{+} ions (having the maximum value of $\mu/Z = 64$, i.e., the maximum time of flight), we can determine the starting instant t_1 for these ions, which is marked by arrows in Figs. 2(a) and (b). The values of μ/Z for the ions generating all other peaks can be determined from the delays of these peaks relative to t_1. The obtained values $\mu/Z= 32, 21$, etc., can be naturally ascribed to Cu^{2+}, Cu^{3+}, etc., ions. This means that ions in all charge states are produced simultaneously. The instant of ion emission was found to be close to the instant of the maximum current rise rate. It can be seen from Fig. 2(b) that a noticeable peak of H^{+} ions was also registered in experiments besides the peaks associated with ions of the cathode material. In addition, ions of other light impurities desorbed from the surface of the cathode and the insulator insert, O^{n+}-, C^{n+}-, N^{n+}- $(n = 1, 2)$, and having values of μ/Z close to those for corresponding copper ions (e.g., Cu^{4+}, Cu^{5+}, Cu^{8+} ions, etc.) may also contribute to the signals. An analysis of the obtained results proved that such ions make an insignificant contribution to the charge composition of the beam. Ion signals detected in various discharges for fixed discharge parameters had a considerable spread in amplitude; for this reason, these signals were averaged for subsequent processing over a series of 12 discharges. The averaged amplitudes of the signals corresponding to ions of a given charge state and measured for various values of E/Z were used for constructing the energy spectrum of such ions.

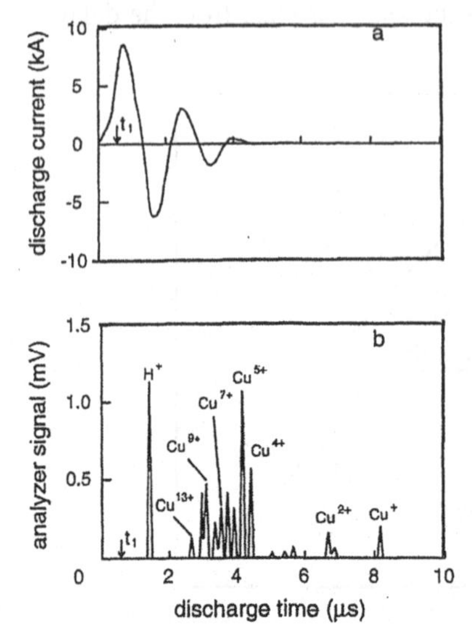

Figure 2. Discharge current at the initial capacitor voltage $U = 2.0$ kV (a) and correspondent signal of the ion energy analyzer with the specified bias voltage ΔU=1.5 kV (b).

Figure 3(a) shows the ion spectra obtained in this way for different charge states of Cu^{n+} for two values of the storage voltage U_0. It can be seen that a typical spectrum consists of a core in which the main fraction of ions with energies not exceeding a few kiloelectronvolts is concentrated and a 'tail' of accelerated ions whose fraction is a few percent of the total number of particles and whose energy can be as high as 150 keV. Another peculiarity observed in Fig. 3(a) is a nearly linear increase in the maximum energy of the ions being registered with their charge. In addition, a comparison of spectra 1 and 2 obtained for different values of storage voltage U_0 shows that the spectra of all ion components are broadened towards higher energies upon an increase in this voltage.

In order to find out whether the acceleration process depends on the ion mass, we made similar measurements with a tantalum cathode. In these experiments, we observed Ta^{n+} ions whose charge composition turned out to be much wider than in the case of Cu^{n+} ions. Tantalum ions in various charge states were also produced simultaneously at the initial stage of the discharge; the duration of this process was found to be close to the corresponding value for Cu^{n+} ions. Using the procedure of signal processing described above, we constructed the energy spectra of Ta^{n+} ions at different storage voltages.

These spectra are shown in Fig. 3(b). It follows that the spectra for tantalum and copper ions have similar structures. The maximum detected energy of tantalum ions also increases with the charge state almost linearly, attaining a value of 350 keV for Ta^{+32} ions at a storage voltage U_0 = 1.5 kV.

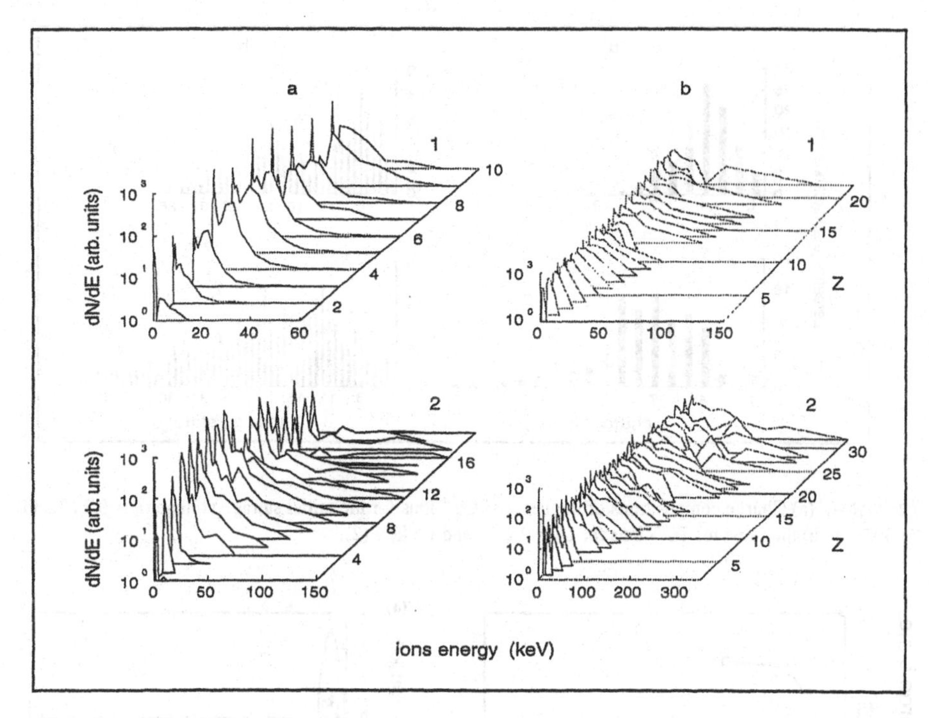

Figure 3. (a) Energy spectra of Cu^{n+} ions for a storage voltages of 1.0 *(1)* and 2.5 keV (2); (b) spectra of Ta^{n+} ions for voltages of 0.2 (1) and 1.5 kV (2).

By integrating over the energy spectra, we obtained the charge state distribution of ion beams of tantalum and copper for different values of storage voltage (Fig.4). It was mentioned above that the resolution of the method used does not allow us to single out the signals of tantalum ions with charges differing by unity in each discharge in the range of high degree of ionization (Z > +30). For this reason, the charge state distribution of the ion beam in this region was plotted not for an individual component, but for groups of unresolved components, and demonstrates only the general tendency of ion distribution. It can be seen from the figure that the charge state similar, but the maximum detectable charge of copper ions is +19, while for tantalum it lies near +50. In addition, it can be seen that the range of the charge states of the ions being detected becomes wider upon an increase in the capacitor voltage.

The dependence of this quantity on the storage voltage is shown in Fig. 5 for both species of ions.

It can be seen that the mean charge of the copper ion beam increases monotonically with U_0, while the mean charge of tantalum ions attains its maximum value for $U_0 = 500$ V and then remains virtually unchanged.

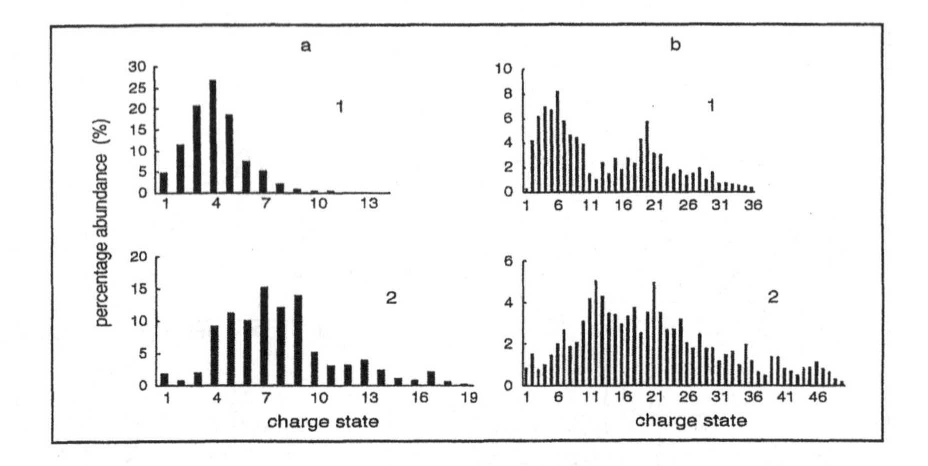

Figure 4. (a) Charge compositions of the beam of Cu^{n+} ions for the initial storage voltage $U_0 = 0.3$ *(1)* and 2.0 kV (2); (b) the same for Ta^{n+}- ions for $U_0 = 0.2$ *(1)* and 1.5 keV (2).

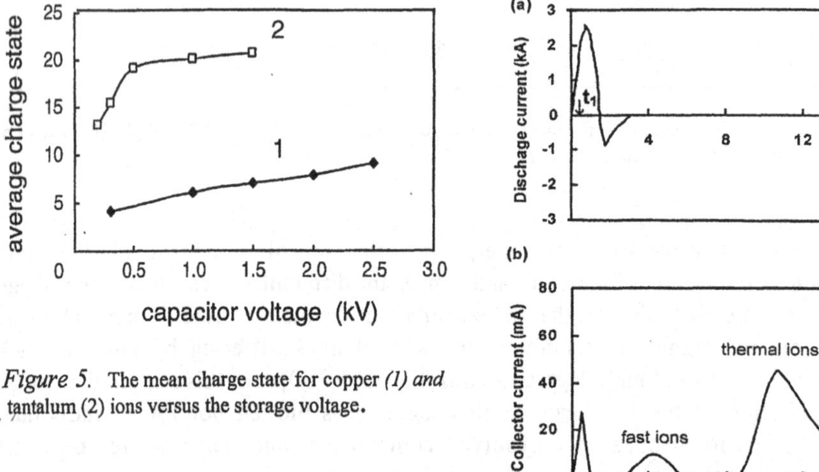

Figure 5. The mean charge state for copper *(1) and* tantalum (2) ions versus the storage voltage.

Figure 6. Oscillograms of the discharge current (a) and collector current (b)

2.2 COLLECTOR MEASUREMENTS OF ION CURRENT

The results presented above demonstrate that the highly charged ions of the cathode material are generated at the initial stage of burning of the vacuum discharge. To estimate amount of these ions the direct measurements of ion current of cathode jet have been performed. Ion current was measured with a collector that was placed at the end of a drift tube of 30 cm length. The collector registered ions, which have passed through the anode and the tube. The typical ion signal is presented in Fig.6(b). It shows that there are two groups of ions in the cathode plasma jet. The first are the 'fast' ions, which approach the collector approximately in 4 μs after the discharge onset. The second, main group of ions, that was called as the 'thermal' ions has the peak of intensity in 11 μs after the discharge onset. We suggest that the 'fast' ions present just the highly charged and high energy ions, which was studied at the previous section. Hence, they have been generated at the same moment, that is pointed with an arrow in Fig.6(a). From delay between the maximum of the signal corresponding to the 'fast' ions and the moment of their generation the mean velocity of the ions was calculated.

With a similar method a mean velocity of bulk of the 'thermal' ions was estimated, however, it was more naturally to refer the moment of their generation to the peak of the discharge current. These estimations were performed throughout the range of variation of the discharge current. The results are presented in Fig.7. Also the mean energy of the highly charged ions, which were considered in the previous Section, is presented in this figure. It was estimated by formula $E \approx e\langle Z\rangle U_0$ from the mean charge state of the ions $\langle Z\rangle$ presented in Fig.5 for given capacitor voltage U_0 that determine the peak discharge current. Fig.7 shows that there is a good agreement throughout the range of measurements between values of energy for the 'fast' ions in the collector measurements and the highly charged ions that have been studied by the energy analyzer. This supports our suggestion on identity of these two types of ions.

So, from the data presented in Fig.6 one can estimate the ion currents for both the 'fast' and the 'thermal' ions. These measurements have been performed also throughout the range of variation of the capacitor voltage. The results show that current of the ' fast' ions attains 200 mA, hence taking into account that value of the collector area of about 7 cm^2 one can derive the maximum density of current for the highly charged ions, it attains 30 mA/cm^2 at a range of 30 cm from cathode.

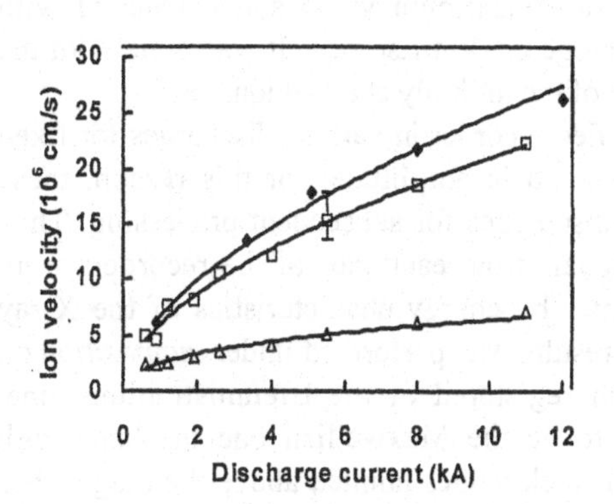

Figure 7. Velocities of multiply charged ions (top curve), 'fast' (middle curve), and 'slow' ions (bottom curve).

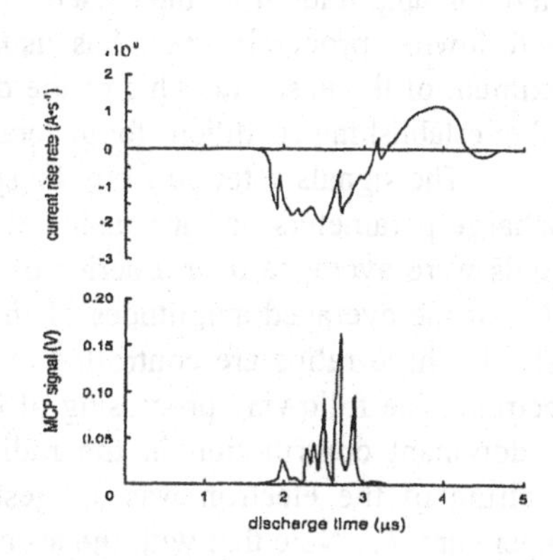

Figure 8. Oscillograms of discharge current rate of rise (top) and X-ray detector signal (bottom).

2.3 . X-RAY MEASUREMENTS.

The above presented studies have demonstrated that a short-run ejection of beams of highly charged ions of the cathode matter occurs from cathode jet of a low – voltage vacuum spark. This result suggests that there arise a local region of highly heated plasma in the cathode jet and the plasma is supposed to emit an X-ray. Hence, the measurements of X-ray emission from the cathode jet of the spark have been also performed by means of a specially designed recorder with a micro-channel plate (MCP) as a detector. MCP is available to register an X-ray emission within a range of energy of 0.01 – 1.0 keV with a yield of about 0.1 electron per quant and the value is near constant at the pointed energy range. An important characteristic of the MCP is its insensitivity to a visible light that allows applying a set of transparent polypropylene films as the absorbers for analyses of an X-ray spectrum. In this experiments three recorders were applied, each of ones was covered with a specified polypropylene film of 3.7, 7.4, 11.1 μm thick. The films kept the plasma particles and ultraviolet radiation from entering the MCP and they transmitted the X-ray of the specified energy ranges. After passing through the absorbers the X-ray radiation entered the MCP, converted into electron flow, the last was magnified by about 10^4 times and entered the corresponding collector placed behind the MCP.

Fig.8 shows oscillograms of signal from the Rogovsky coil and a typical signal from one of the recorders. One can see that at the initial stage of the discharge burning near the peak of the current rise rate an X-ray emission from the discharge plasma was registered. The emission presents a sequence of 'flashes, each of about 100 ns length (FWHH), which correlates with the peculiarities at signal of discharge current rise rate. By comparison of Fig. 8 with Fig.2 one can see that interval of the X-ray emission correlates with the time of the highly charged ions emission. The signals that have been recorded at the same discharge from the other registrators have the same shape but they differed in amplitude from the presented one. A peak at the oscillogram was adopted for the following processing that has usually the maximum value and correlated with maximum of the absolute value of the discharge current rise rate. It was done with the goal of establishing conditions for generation of the multiply charged ions.

The signals detected from the specified recorder in various discharges for fixed discharge parameters had a considerable spread in amplitude; for this reason, these signals were averaged over a series of 12 discharges for subsequent processing. Then ratios of the averaged amplitudes of the signals from each pair of the recorders were derived. These ratios are controlled evidently by energy characteristics of the X-ray spectrum. The following processing of the results was performed under *supposition* on the dominant contribution in the radiation registered of the Bremmstrahlung one. Spectrum of the electron was suggested to be the Maxwellian one in shape with temperature T_e. Note that with the absorber thick values pointed above the diagnostics technique provides the relevant accuracy (about 20 %) at the range of temperatures T_e= 0.1 – 0.6 keV. The temperature obtained in such a way lies at the range of 150 – 250 eV throughout the range of variation of the discharge parameters and tends to increase with the capacitor voltage (see Fig.9). The temperatures derived from two pairs of the recorders were closed to each other near throughout the range of measurements that supports the assumption on the Bremmsrahlung nature of the detected radiation.

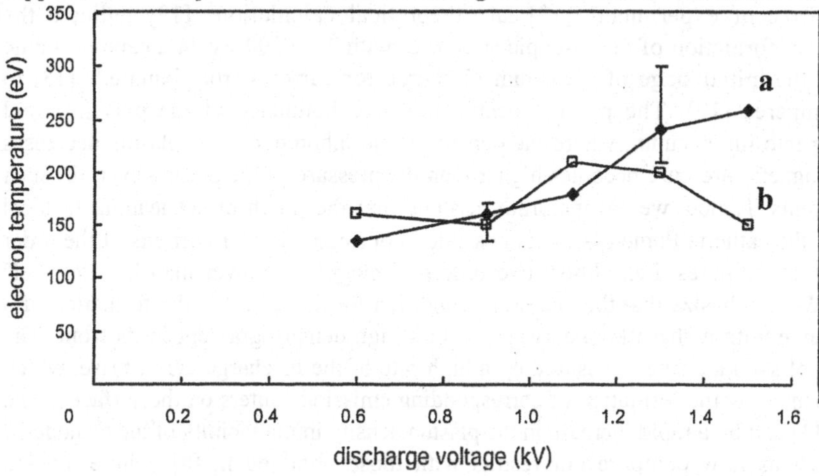

Figure 9. Dependence of the electron temperature obtained from pair of signals of the X-ray recorders covered with absorbers of (a) 3.7/7.4 μm and (b) 7.4/11.1 μm thickness.

Discussion

The results of measurements presented in Section 2 show that the plasma jet of a vacuum discharge produces beams of accelerated multiply charged ions of the cathode material. The obtained values of the ion charge are much higher than those observed earlier in various types of vacuum spark. Multiply charged ions can be produced only in plasma with a high electron temperature. The results of the X-ray measurements show that, in fact, at the moment of generation of the ions the 'flashes' of X-ray radiation are registered. These 'flashes' are explained by formation at plasma of the cathode jet of a local region, where electron temperature attains 150 – 250 eV. Hence, results of the X-ray measurements support results of the ion measurements.

Now let us consider a possible mechanism for production of this region. An important feature of the production of multiply charged ions is the short-term "burst" nature of the process. Ions are produced at the initial stage of the discharge 400-600 ns prior to the attainment of the peak value of current. The duration of this process is several tens of nanoseconds and is much smaller than the time of the discharge current rise. This indicates a small spatial scale of the ion production region, which can be estimated using the measured value of $\delta \geq \Delta t.V = 6 \times 10^{-2}$ cm (here, V = 1.2 x 10^6 cm/s is the velocity of the cathode jet in spark discharges [11]). It follows hence that multiply charged ions are produced in a local region of hot plasma in the front of the cathode flame moving towards the anode, i.e., before the plasma fills the discharge gap.

The reason for the formation of the local region in the hot plasma may be compression by the intrinsic magnetic field of the current. This phenomenon is well known for high-current discharges for which the formation of such microscopic regions at the constriction of the plasma filament due to the pinch effect was observed at a late stage of the discharge for currents of the order of 100 kA and higher [5,6]. However, the results of recent experiments [12] and theoretical calculations [13] indicate the possibility of formation of microscopic structures with $T_e \sim 100$ eV in a cathode flame plasma at the initial stage of a vacuum discharge for currents from hundreds [13] to several amperes [12]. The process occurs near the boundary of the plasma cloud expanding into the vacuum, where the density of the inhomogeneous plasma decreases and the magnetic pressure becomes higher than the pressure in the plasma even for such weak currents. Hence, we can naturally assume that the pinch effect leading to local heating of the cathode flame plasma is possible in our case also for currents of the order of several kiloamperes, i.e., almost two orders of magnitude lower than in "classical" systems. We emphasize that the necessary condition for this effect is the formation of a plasma flame with a sharp boundary, i.e. with a high density gradient in its front. The formation of such a flame is ensured by a high rate of the discharge current rise, which is accompanied by the formation of corresponding emission centers on the surface of the cathode [11] and by a rapid increase in the plasma density in the vicinity of the cathode.

Let us now compare our results with those obtained in [8], where the ion compositions of the plasma of a vacuum spark with currents close to values typical of our experiments (about 10 kA) was investigated. However, in contrast to our experiments, the measurements were made for time intervals considerably exceeding the time of filling of the discharge gap with the plasma and led to the maximum charge +7 for copper ions, which is much lower than the values obtained by us. This indicates the fundamental role of the inhomogeneous structure of the cathode jet in the production of multiply charged ions at the initial stage of the discharge.

Also compare our results with the parameters of the ions of the plasma formed as a result of exposure of the target to a high-power laser pulse. The measurements of the energy and charge distributions of ions of the laser plasma were made using a technique similar to that used in our experiments. Ion beams in the experiment described above and the ions of laser plasma have similar characteristics. For example, the charge state distribution and the form of its variation are the same for different materials of the cathode (target), and the maximum detected values of charge and energy of the ions of a given species are close. Also we see that current density for the multiply charged ions in our source succeeds the corresponding values at the laser-produced plasma sources. A structure of the energy spectra of ions similar to that depicted in Fig. 3 was observed for multiply charged ions of a laser plasma [3]. As in our experiments, the maximum and mean energies of different ion components were almost linear functions of their charge states.

We sum up the results of the consideration presented above in Table 1 as follows:

It has been shown that the charge state distribution and the energy distribution of ions are close to the parameters observed in a laser plasma, while the typical values of supplied energy are almost two orders of magnitude lower. This indicates that a spark discharge can be used in principle as an effective source of accelerated multiply charged metal ions.

TABLE 1. Parameters of ion beams produced from different types of sources

Parameters of beam of multiply charged ions	Type of ion source		
	Steady-state or 'slow' (~1ms)	'Fast' ($\leq 1\ \mu s$)	
		Laser produced plasma	Low – energy vacuum spark
1. Current density, *mA/cm*^{2}	$10^2 - 10^3$	10 – 20 (at 90 cm)	up to 30 (at 30 cm)
2. Mean charge state	3.4 (Pb) 3.0 (Cu) 4.3 (Ta)	41 (Pb) 15 (Cu) 42 (Ta)	9.3 (Cu) ~ 20 (Ta)
3. Maximum charge state	+7 (Pb) +4 (Cu) +6 (Ta)	+49 (Pb) +25 (Cu) +55 (Ta)	+19(Cu) ~+50 (Ta)
4. Maximum energy, *MeV*	------	1 – 4	0.15 –0.4
5. Electron temperature, *keV*	--------	0.5 – 1.5	0.15 – 0.25

References

1. A. Bugaev, V. Gushenets, G. Yushkov, E. Oks, T. Kulevoy, A. Hershkovitch and B.M. Johnson, *Appl. Phys. Letters* 79, 919 (2001)
2. A. Anders, G. Yushkov, E. Oks, A. Nikolaev and I. Brown, *Rev. Sci. Insrum.* **69**, 1332 (1998)
3. W. Mroz, P. Paris, J. Wolowski, W.Woryna , *Fusion Eng. Design*. **32**, 425 (1996).
4. L. Laska, B. Kralikova, J. Krasa, K.Masek, *Proc.13th LIRPP'97, Monterey, USA, April 13-18* , 1 (1997)
5. K. N. Koshelev and N. R. Pereira, *J. Appl. Phys.* **69**, R21 (1991).
6. C. R. Negus and N. J. Peacock, *J. Phys. D* 12, 91 (1979).
7. A. A. Gorbunov, M. A. Gulin, A. N. Dolgov, *et al.*. Preprint No. 024-88, MIFI (Moscow Engineering Physics Institute, Moscow, 1988).
8. A. Anders, I. G. Brown and R. A. MacGill,, *IEEE Trans. Plasma Sci.* **25**, 718 (1997).
9. E. D. Korop and A. A. Plyutto, *Zh. Tekh. Fiz.* **37**, 72 (1970) [Sov. Phys. *Tech. Phys.* **12**, 53 (1970)].
10. M. P. Stockli and D. Fry, *Rev. Sci. Instrum.* **68**, 3053 (1997).
11. G. A. Mesyats, *Ectons* (Nauka, Yekaterinburg, 1994), Part 2.
12. Nadja Vogel, *Proc. 18th ISDEIV, Eindhoven , the Netherland,* 202 (1998)
13. E. A. Zverev and I. A. Krinberg, Pis'ma Zh. Tekh. Fiz. **24**, 486 (1998)

STATUS OF MEVVA EXPERIMENTS IN ITEP*

Kulevoy T.V., Kuibeda R.P., Pershin V.A.,
Batalin V.A., Seleznev D.N., Petrenko S.V.,
Zubovsky V.P., Kolomiets A.A.
Department of linear accelerators
Institute for Theoretical and Experimental Physics
Moscow
B.Cheremushkinskaya 25, 117259, Russia.

Abstract. The MEVVA ion source and different its modification are investigated in ITEP for a long time. The last results obtained in the experiments with e-MEVVA and MEVVA-M (MEVVA with two anodes) ion sources are presented. As well, the common MEVVA is used as injector for RFQ linac providing experiments with plasma target. The results of copper ion beam acceleration are presented.

1. Introduction

The MEVVA ion source (IS) is developed in Moscow Institute for Theoretical and Experimental Physics as ion beam generator for heavy ion accelerators. The interest to this ion source was initiated in 1986 year in frame of Heavy Ion Inertial Fusion (HIIF) program. For this program the high current (about tens mA) beam of ions with mass larger than 200 a.m.u. was needed. Exactly at that time, the Ian Brawn papers appeared [1] and the MEVVA IS started his life as ion source for accelerators. Since that time the MEVVA is one of the best candidate as ion source for HIIF driver. Now in ITEP the MEVVA IS is used as the injector for heavy ion linac HIPr-1 (Heavy Ion Prototype). At the linac the set of experiments with plasma target are prepared in collaboration with our institute plasma physics group. Also the investigation of different modification of MEVVA IS, those can result the increasing both the average and maximum ion charge state in high current beams, are going on. More than ten years ago in ITEP the scientific group of d. Batalin suggested the idea of e-MEVVA ion source [2]. In this IS the accelerated high-density e-beam is used to enhance the charge state of ions generated by MEVVA. Last year in collaboration with Brookhaven Laboratory USA (scientific group of Ady Hershkovich) and HCEI Tomsk (scientific group of Efim Oks), the experimental proving of the idea was obtained [3,4]. During the experiments with e-MEVVA the new

E. Oks and I. Brown (eds.),
Emerging Applications of Vacuum-Arc-Produced Plasma, Ion and Electron Beams, 39–49.

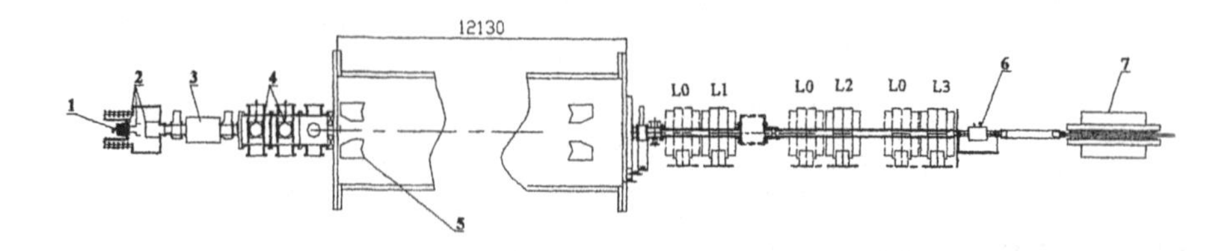

Figure 1 Experimental layouts for plasma target experiments at HIPr-1.
1 – MEVVA IS; 2 – Extraction system; 3 – Measurement chamber; 4 – electrostatic lenses; 5 – RFQ; 6 – plasma target; 7 – MA.

effect that provides enhance of beam ions charge states without e-beam was found. It was shown that the using of the two anodes with staggered high current discharge in axial high magnetic field gradients results significant increasing of ions charge state. The beam with significant part of uranium ions with 6+ and 7+ charge states was generated.

The new test-bench for MEVVA IS investigation is under construction. The two charge state distribution (CSD) measurement systems are installed at that test bench – time-of-flight and magnetic mass analyzer. The equipment for emittance measurements with CCD camera and the electrostatic deflector for beam energy spread measurements are about ready to be installed at the test-bench. We plan that in nearest future we find possibility to use at the test bench equipment providing direct measurements of MEVVA plasma parameters.

2. Plasma target experiments

In ITEP the linac HIPr-1 (Heavy Ion Prototype for HIIF program) that used the MEVVA ion source as the injector was modified few years ago [5]. The output energy of accelerated beam was increased up to 110 keV/nucleon (previously 36 keV/nucleon). Linac provides acceleration of ions with specific mass till 60 a.m.u. (like U^{4+}) with current till 12 mA. Now the MEVVA IS with copper cathode is used at this linac as the ion beam injector. The accelerated beam of Cu^{2+} ions with current of 6 mA has been obtained. Now the set of experiments with plasma target are under preparation at this linac. There are several scientific experiments which will be done using the beam provided by the linac. Those are: i) the measurements of energy loss in a thin plasma target as well as the mean effective charge state – parameter that used for description beam-plasma interaction. ii) The measurements of charge state distribution changing into the plasma target for ions of the same material but with different charge states. iii) The measurements of cross sections for both beam-plasma and beam-gas interaction. The plasma target created by electric discharge in the hydrogen gas was carefully tested and was used for a set of experiments with a proton beam [6]. At HIPr-1, all experiments will be started with copper beam but the set of another materials (at last the uranium beam) are planed as well. At the beginning, the beam from common MEVVA will be used. As soon as the experimental layout is tuned and first results are carried out, the e-MEVVA and MEVVA-M IS will be installed at the linac. The experiments with U^{7+} ion beam promise the most interest. The experimental layout is shown in Fig.1. To match the ion beam generated by MEVVA IS with accelerating RFQ structure, the two electrostatic lenses are used. At the output of accelerator, three quadruple lenses (L1, L2 and L3) provide the accelerated beam focusing on plasma target. It is necessary to have the ion beam with current of 100 micro Amps into plasma target input diaphragm with diameter of 1 mm. The simulation shows that to provide this current into plasma target, the accelerated beam should be at least 2 mA. The accelerated Cu^{2+} beam current up to 6 mA has been already obtained.

3. E-MEVVA experiment

The idea of e-MEEVA ion source consisted of combining an electron beam, a vacuum arc ion source, and a plasma drift tube, to provide the charge states enhance of generated ions was suggested in ITEP more than ten years ago. The first encouraging indications of higher charge state production was reported at several conferences in 1994 year but results were not convincible. The international scientific collaboration gave new life for this idea and finally last year the first undoubted experimental proof was received as a result of ITEP-HCEI-BNL collaboration.

The new plasma e-gun with 23 cm e-beam drift channel was constructed in Tomsk for these experiments. In ITEP the new version of MEVVA plasma generator, new coils for plasma drift chamber and several power supplies for e-gun and coils were constructed. In addition, the separate e-gun High Voltage platform (eHVp) was constructed also.

For first experiments, the lead cathode with axial hole of 4 mm was used. The dimension of cathode has been taken from previous experience. However it was found that the hole has to be enlarged at least up to 11 mm.

The highest vacuum in experimental tank, which e-MEVVA is connected to, that could be obtained, when this version of e-gun is installed, is $1*10^{-6}$ mBarr. The stable operation of the e-gun could be provided when vacuum in the experimental tank is better than $1.6*10^{-6}$ mBarr. The long drift tube (~600 mm) with small diameter (36 mm) as well as 4 mm hole in MEVVA cathode reduce the vacuum condition and pumping possibility of the e-gun region. As a result, it was possible to provide the operation mode of e-gun (recommended by Tomsk) only for 2-3 hours. During this time the conditioning of the e-gun took at least one hour. The highest emission e-beam current was 10 A.

It is not enough for complete CSD measurement of ion beam generated by e-MEVVA with analyzing magnet. Nevertheless, it was possible to check the feature of separate peaks of interest. Such experiments have been done. It was found that amplitudes of following peaks – H^+, O^+, Pb^{2+}, Pb^{3+}, Pb^{4+} – didn't change their value for both mode (with/without e-beam). It was concluded that this version of e-gun didn't provide the e-beam with high enough density for increasing the charge states of ion beam generated by e-MEVVA ion source.

Taking into account these experimental facts, the new MEVVA cathode support flange for cathode with 12 mm hole has been designed. Also several electrodes of

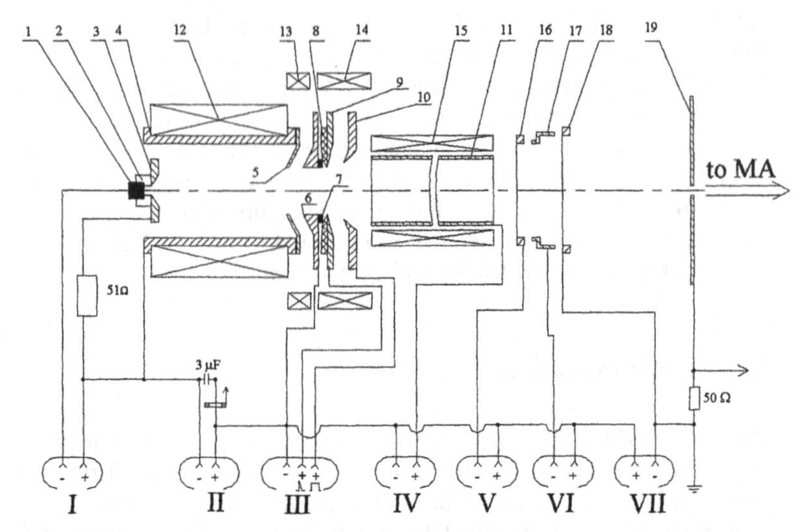

Figure 2. ITEP e-MEVVA with Tomsk e-gun. 1 – e-gun cathode, 2 – e-gun insulator, 3 – e-gun trigger, 4 – e-gun anode, 5 – e-gun plasma grid, 6 – e-beam extracting electrode, 7 MEVVA cathode, 8 – MEVVA insulator, 9 – MEVVA trigger, 10 MEVVA anode, 11 – plasma drift channel, 12 – e-gun anode coil, 13 – transport coil, 14 – MEVVA coil, 15 – PDC coil, 16 – ion beam extractor, 17 – focusing electrode, 18 – ground electrode, 19 – first slit at the input of MA. I – Trigger pulse for e-gun, II – accelerating e-beam voltage, III – MEVVA discharge pulse voltages, IV – potential at the PDC, V – ion extraction voltage, VI – ion beam focusing voltage

MEVVA were modernized. As well it was concluded that it is necessary to have MEVVA discharge aria as near as possible to e-beam extraction gap in e-gun. For this goal, the Tomsk team sent us the new plasma flange for e-gun and we constructed the new insulator flange to adjust this new construction to our MEVVA part of e-MEVVA. The ITEP e-MEVVA with Tomsk e-gun is shown in. Fig.2. The gap between the e-beam extracting electrode and MEVVA discharge atria in new assembling is 30 mm instead of 230 mm in previous experiments.

The new assembling immediately results improving of the vacuum. The vacuum of $5*10^{-7}$ mBarr in experimental tank has been obtained after two days of pumping. During experiments, the vacuum in experimental tank always was better than $8*10^{-7}$ mBarr. Also the conditioning of e-gun takes only a half of hour. As a result, the stable operation of e-gun can be supported for any desirable time. The maximum emission current of 25 A under 18 kV of e-beam accelerating voltage has been reached.

It is a place to note some difference of ITEP e-MEVVA from the HCEI Tomsk one. The main is a design of the plasma drift channel (PDC). The ITEP's one mechanically and electrically separated from MEVVA anode meanwhile in Tomsk version the MEVVA long anode used as PDC. The metal pipe of PDC in ITEP e-MEVVA can be connected electrically to anode of MEVVA part, but for most experiments, it was fed by separate power supply. The total length of ITEP PDC is about 70 cm and diameter is 3.6 cm instead of 40 cm length and 4 cm diameter for the Tomsk's one. Therefore, to provide the same vacuum condition into e-gun region, more power pumping should be provided for ITEP e-MEVVA compare to Tomsk's one. From other hand, the longer PDC is used, the larger the time of interaction between plasma ions and e-beam.

Therefore, to obtain the same $j\tau$ coefficient for increasing of ion charge states in ITEP e-MEVVA, the e-beam density can be smaller than in the Tomsk e-MEVVA.

The maximum emission current and the most stable operation condition have been obtained under following parameters of e-MEVVA (Table 1).

Table 1. e-MEVVA parameters

Parameter	Value
Discharge current at e-gun	100 A
e-beam accelerating voltage	18 kV
Magnetic field in e-beam extraction gap	0.5 kG
Discharge current in MEVVA	100 A
Magnetic field in plasma drift channel	0.5 kG
Ion beam extracting voltage	20 kV

The increasing of magnetic field either in plasma drift channel or in e-beam accelerating gap or in both coils together immediately gives to breakdown. Such low magnetic field along ion source results low ion beam extracted from ion source. It provides some hardships for experiments that will be described below.

The accelerated e-beam has been detected at the output of ion source when the ion beam optic has been off. The Faraday cup is placed at the distance of about 1 meter besides the last grid electrode of ion beam optic. Nevertheless, the e-beam current of 20 mA has been detected. If the MEVVA discharge has been on, the e-beam current has been 120 mA. The several possible explanation of such e-beam current increasing at the FC can be suggested. First, the plasma of MEVVA discharge moves to the e-gun and provides the "flat" accelerating e-beam electrode, meanwhile without this plasma the e-beam extraction electrode has a large hole and a large radial component of extraction field. Second, the plasma of MEVVA works as the plasma lens for e-beam. Third, the plasma of MEVVA compensates the charge of e-beam. It looks like all of them take place and improve transmission of e-beam throughout of ion source.

The ion beams without e-beam and with e-beam are shown in Fig.3. When the e-beam is on, it is easy to see influence of it on the ion beam. The ion beam has some valley during e-beam and an increase of amplitude after e-beam. Exactly this part of ion beam contains the ions with increased charge states.

The measured charge state distributions of lead ion beam generated by e-MEVVA with/without e-beam are shown in Fig.4. For "without e-beam" mode the maximum

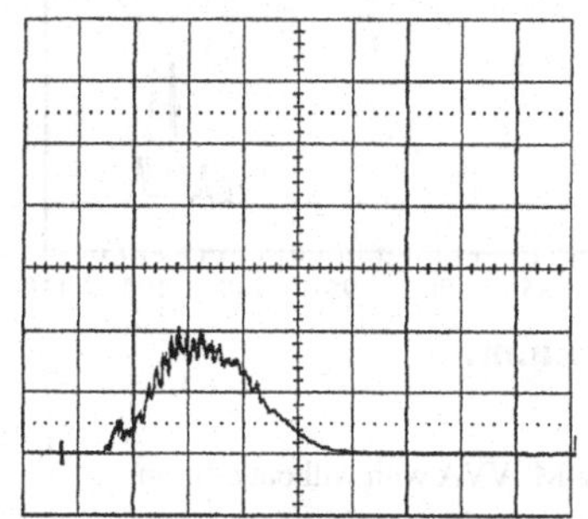

Figure 3a.. Ion beam at the output of e-MEVVA without e-beam. 10 mA/dev. 50 μs/dev

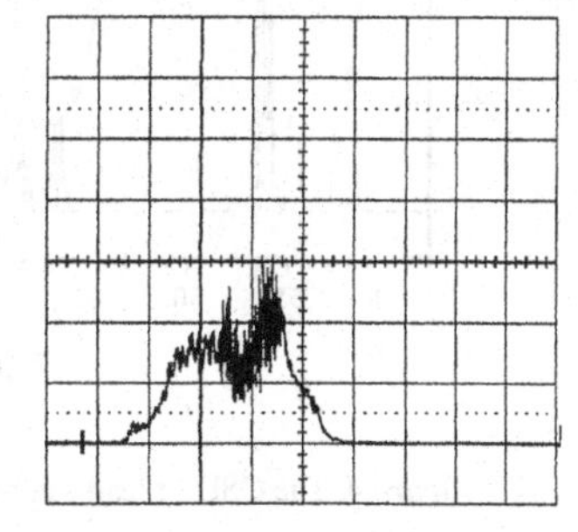

Figure 3b. Ion beam at the output of e-MEVVA with e-beam. 10 mA/dev. 50 μs/dev.

amplitude of signal was measured independently of it's time position along measured pulse. For "with e-beam" mode, the maximum amplitude of signal during and after e-beam injection time was measured only. Such method explains the large difference in amplitude of peaks corresponding to residual gas ions (like H^+, C^+, O^+). The main part of them is at the beginning of the ion current pulse and for "without e-beam" mode provides large amplitude of signal. The middle and end of beam pulse usually contain significantly less such kind of ions.

The one can see that e-beam significantly decreases the amplitude of Pb^{2+} peak (about five times). From other side, the e-beam increases the amplitude of Pb^{4+} peak. The amplitude of this peak increases about in one order. The peak of Pb^{3+} ions keep the same amplitude both for with e-beam and without e-beam mode. Therefore we can assert that e-beam increase the average charge of ion beam generated by ion source. It is impossible to say something definitive about generation of Pb^{5+} ions. As it was mentioned above, to transport e-beam to MEVVA and plasma drift tube region, we had to use the relatively low magnetic field along these regions. As result the generated ion beam has amplitude too low to provide measurements at the magnet analyzer. To increase the sensitivity of magnet analyzer, we had to reduce the resolution of measuring system. The identification of Pb^{2+} - Pb^{4+} ions did not make problem, but Pb^{5+} ions peak could not be separated from next ones. Also small sensitivity of MA explains the absence of charge-exchange peaks in CSD. Such kind of peaks usually has amplitude one order less than the "original" ones.

The presence of Fe^+ and Zn^+ ions in CSD for "with e-beam" mode indicates that e-beam touches the surface of ion source.

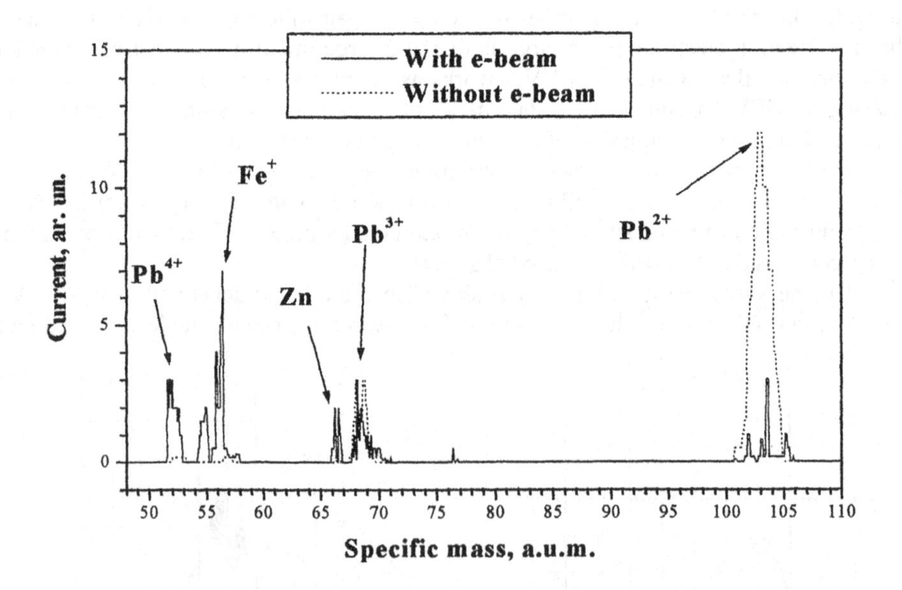

Figure 4. The CSD of lead ion beam generated by e-MEVVA with/without e-beam

4. MEVVA-M experiment

During experiments with e-MEVVA ion source it was found that this ion source can generate the ion beam with higher charge states than the common one when the external e-beam is off. To provide this effect, the two anode step-by-step high current discharge and axial high gradient magnetic field were used

The ion source design and its electrical layout are shown in. Fig.5. To trigger the discharge, as for usual MEVVA, the high voltage pulse is applied between cathode and trigger electrode separated by ceramic insulator. The plasma plume reaches the anode-1 and the discharge between cathode and this anode initiates. The discharge current in this circuit is 120 A with pulse length 150 μs. Then the plasma plume reaches the anode-2 and discharge in circuit cathode – anode-2 starts. The discharge currents in the circuit cathode – anode-2 800 A and 2400 A were used during experiments. Discharge plasma drifts inside the anode-2, which length is 64 cm, to the grid extraction system. To keep plasma near the source axis during this drift, the axial magnetic field B_{a2} provided by coil SDC {Solenoid of Drift Channel) is used. The amplitude of this magnetic field always was 1 T. The coil SeM (solenoid of e-MEVVA) provides the axial magnetic field at the main discharge aria. It can be varied in range from 0 (when it is off) to 0.2 T.

To measure the ion beam CSD, the magnetic analyzer is used. During measurements, the maximum current value of detected beam at the output of analyzer was taken for each step of magnetic field. That method gives the overstating value for peaks of residual gas ions. These ions mainly generates at the triggering moment of discharge and present at the beginning of the beam for short time. At the main part of the beam total amplitude of residual gas ions current is less than 10% of total beam current. Anyway, for every measurement set, we used that method and measured all

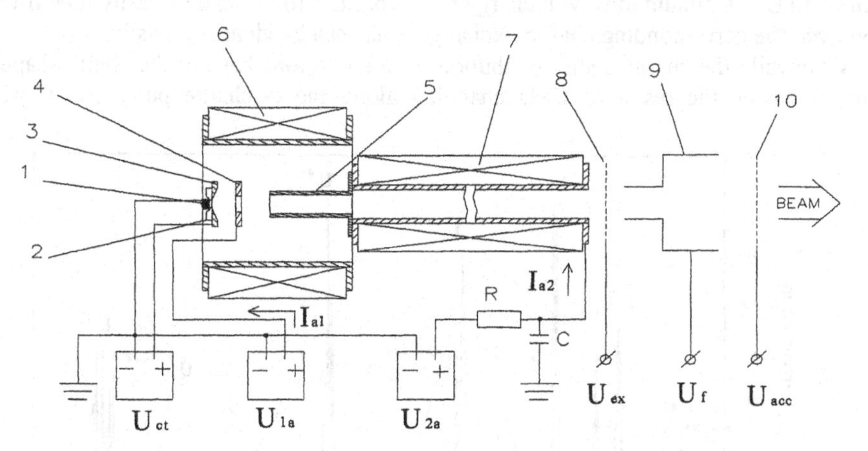

Figure 5. Electrical layout of MEVVA-M. 1 –cathode, 2 –ceramic insulator, 3 – trigger, 4 – anode1, 5 – anode2, 6 – coil SeM, 7 –coil SDC 8 –extractor,9 – focusing electrode, 10 – accelerating electrode, 11 – magnetic mass analyzer.

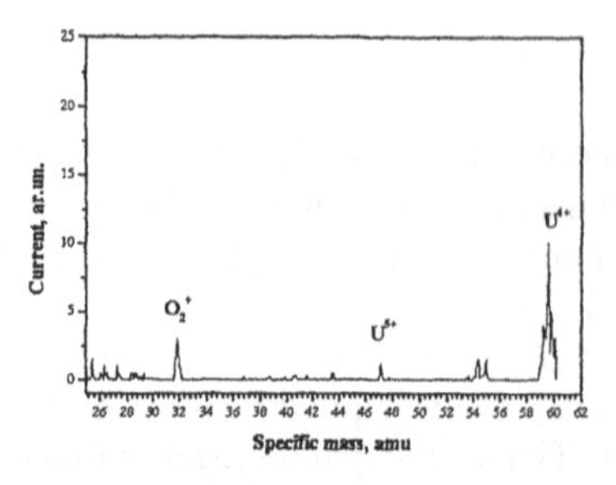

Figure 6a. Uranium beam CSDfor discharge currents 800.$B_1 - 0.2$ T. $B_2 - 1$T

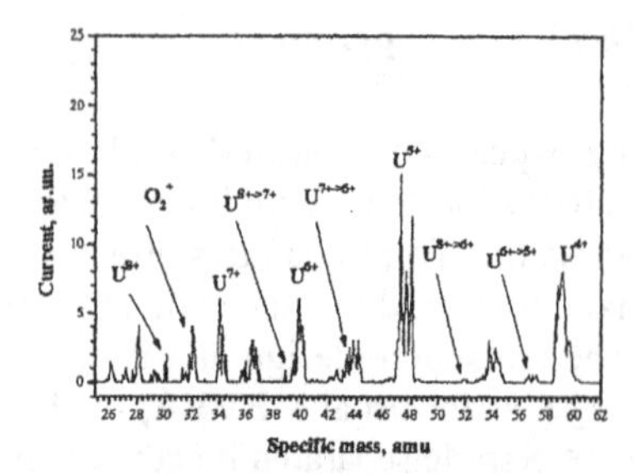

Figure 6b. Uranium beam CSD for discharge currents 800. $B_1 - 0$ T. $B_2 - 1$ T

peaks for range of specific mass from 12 to 80 (include from C^+ to U^{3+}) because the peaks of residual gas are used for interpretation of obtained results. As it was mentioned above the discharge current in circuit anode1-cathode always was 120 A. The magnetic field along the anode2 (drift tube) also was the same (1 T) for all experiments. For different discharge currents in circuit cathode – anode2, the CSD were measured in two modes: when the coil SeM was off and when it provides 0.2 T at its axis.

The results of measurements for discharge currents 800 A are shown in Fig.6. It was found that for cases when the coil SeM is off, the ion source provides higher charge states compare with the mode when both coils are on.

As well it was found, that if the discharge current of 800 A in cathode-anode-2 circuit was used, the generated ion beam has significant part of U^{7+} ions. Also it was found that charge states increased with discharge current. The result of the CSD measurements for discharge current of 2.5 kA in circuit cathode – anode-2 is shown in Fig.7. The measurements have been carried out for mass region corresponding for ions from C^+ to U^{3+}. Uranium ions with charge state from U^{3+} to U^{8+} can be easily identified. Moreover, the corresponding charge-exchange peaks can be identified easily.

Meanwhile the charge state distribution of uranium ions has not the "bell" shaped form. It can be the result of CSD changing along the discharge pulse. As it was

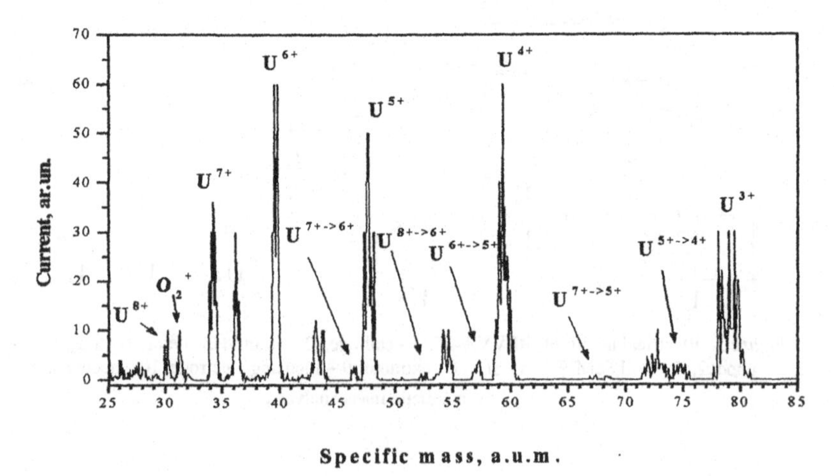

Figure 7. The CSD measurements for discharge current of 2.5 kA. $B_1 - 0$ T. $B_2 - 1$ T..

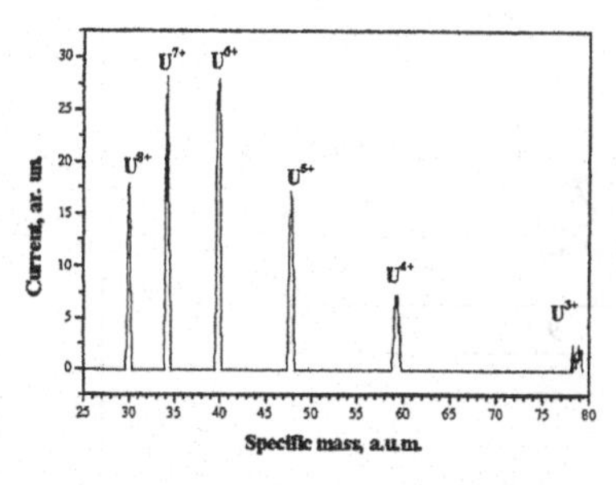

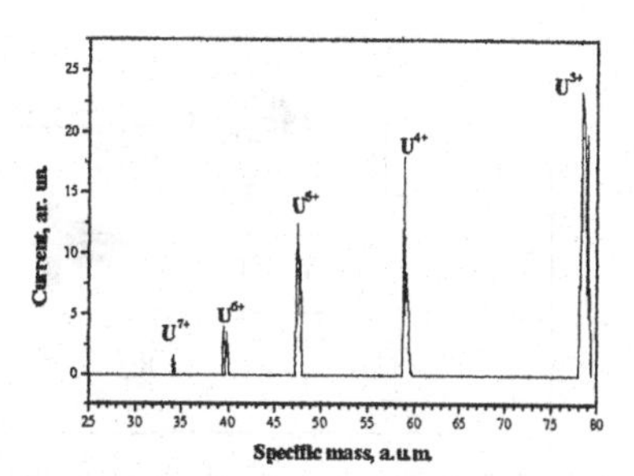

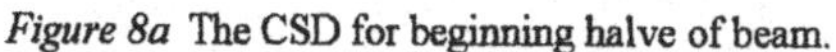

Figure 8a The CSD for beginning halve of beam. *Figure 8b.* The CSD for end halve of beam.

mentioned above, the maximum current was measured for each step of magnetic field independently of its location along beam pulse. Therefore, if the CSD of generated beam changes during the pulse, the maximums for different charge states are separated in time and, as result, the CSD should has the form like it was measured. To investigate this effect, we have carried out the measurements of CSD only for uranium ions with averaging the measurement result for ten pulses per each magnetic field step.

During measurements the beam pulse has been divided on two half and amplitudes have been measured for both half separately. Results are shown in Fig.8. It is necessary to note that for both part of beam we have bell shaped CSD. For beginning of the pulse the maximum corresponds to U^{7+} and U^{6+} ions and for end the CSD looks like one measured for usual MEVVA with high discharge current. Also it is necessary to note that if take this two plots together one will have the CSD with local minimum on the peak of U^{5+} ions like it is on Fig.7. It means that difference of uranium CSD from bell-shaped is a result of changing the CSD from high values in the beginning to smaller ones in the end of beam pulse.

The discharge current for different mode of magnetic field distribution was measured. Simultaneously the voltage at the capacitor C = 200 μF was measured by resistor divider. It was found that for mode with magnetic cork the discharge current reach the higher value than for SeM on mode. As result, the power for the mode "SeM off" (with magnetic cork) is more than 5% higher than for the mode "SeM on".

5. New test-bench

The promising results obtained in experiments with e-MEVVA and MEVVA-M IS call forth the new test-bench creation. It is planned that this test-bench will be used both for testing any IS for using at accelerators of our Institute and for investigation of new ideas in ion source domain. The layout of test-bench is shown in Fig.9. It includes two similar vacuum tanks with insulator flanges which ion source connected to. The equipment for time-of-flight measurements is installed in one tank. For this measurements, the grids accel – diccel system is used for ion beam extraction from IS. To form a short beam pulse, the voltage pulse with magnitude up to 1.5 kV and pulse length of 70 microsecond is applied to electrostatic deflector. The length of drift channel between deflector and beam detector is 1.8 m.

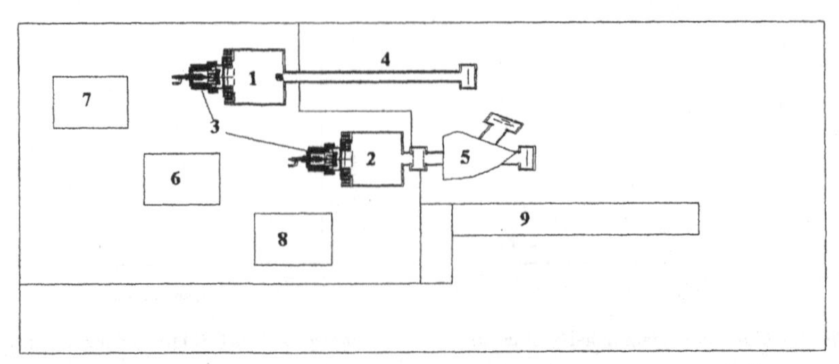

Figure 9. The layout of test-bench 1. ToF tank, 2. MA tank, 3. Ion source, 4. ToF equipment, 5. Magnet, 6. IS HV platform, 7. E-gun platform, 8. Operator system

The magnetic analyzer (MA) that was used for all our experiments with e-MEVVA and MEVVA-M IS was moved from HIPr-1 and connected to second vacuum tank. Main parameters of Magnet Analyzer (MA) are given in Table 2.

Table 2 Parameters of magnet analyzer.

Bending angle	60°
Magnetic field range	0.1 – 1 T
Bending radius	0.3 м
Minimal step of magnetic field changing	$3*10^{-4}$ T
Width of magnetic field	3 см
Magnetic gap length	2 см

For ion beam collimation, two vertical slits of 2 mm width are installed at the input of the analyzer. The distance between them is 180 mm. The experiments with copper beam from common MEVVA showed, that system enables separate isotopes of copper ions – Cu_{63} and Cu_{65}. In order to increase the sensitivity of measured system, the one of this slit can be taken out. As result the resolution of system is reduced but for most important experiments it does not significant. Especially if ion beam of heavy and low charged ions is investigated. At the MA output, the plate located behind the 2 mm slit is used as detector. Between slit and plate, the electrical field accelerating secondary electrons from measuring plate can be applied. It is done to increase the sensitivity of the detector. The coefficient of secondary electron emission can be different for different kind of ions. Nevertheless, this effect is not principle because the main goal of experiments with e-MEVVA and MEVVA-M is to find changing of peak amplitude for definite ions. The one hole extraction system for this measurements is now under design by KOBRA-3D code.

Two high voltage platforms are installed at test-bench. One of them is already under operation and first tune experiments of TOF system with beam from common MEVVA were carried on. When both systems are put under operation and all support jobs is over the e-gun high voltage platform will be moved and installed.

For the ion beam emittance measurements, the needed equipment (paper-pot mask, scintillate, and CCD camera) can be installed in a standard measuring chamber where the diaphragms for MA are placed now. The same chamber with equipment for emmitance measurement can be easily installed at the second vacuum tank instead of drift channel. The chamber with 45° electrostatic deflector for beam energy spread measurements can be installed at any moment at both tanks.

Acknowledgments

Names of A. Hershcovitch, B.M. Johnson (BNL), E.Oks, A. Bugaev and V.Gushenets (HCEI Tomsk) should stay at the title of this report because the main and most successful part of experiments presented in this paper were done only thanks to them. We were glad to work with them and hope that our good collaboration to be continued and gives a lot of new good results.

References:

1. I.B.Brown, J.E.Galvin and R.A.MacGill, *Appl. Phys. Lett.* 1985, 47, 358.
2. Batalin, V.A., Volkov, Y., Kulevoy, T.V. & Petrenko, S.V. (1994), *Proceedings of the 4th European Accelerator Conference* (EPAC-94, London, England) 1560-1563; *Proceedings of the 17th International LINAC Conference* (LINAC-94, KEK, Ibaraki, Japan) 390-392
3. Batalin, V. A., Bugaev, A.S., Gushenets, V.I., Hershcovitch, A., Johnson, B.M., Kolomiets, A.A., Kuibeda, R.P., Kulevoy, T.V., Oks, E.M., Pershin, V.I., Petrenko, S.V., Seleznev, D.N. & Yushkov, G.Yu. (2002a).). Further development of the E-MEVVA ion source. *Rev. Sci. Instrum.* 73, 702-705.
4. Batalin, V. A., Bugaev, A.S., Gushenets, V.I., Hershcovitch, A., Johnson, B.M., Kolomiets, A.A., Kuibeda, R.P., Kulevoy, T.V., Oks, E.M., Pershin, V.I., Petrenko, S.V., Seleznev, D.N. & Yushkov, G.Yu. (2002b). Electron-beam enhancement of the metal-vapor vacuum-arc ion source. Further Development of the E-MEVVA Ion *Journ. Appl. Phys.* (pending).
5. D.Kashimsky, A.Kolomiets, T.Kulevoy, R. Kuibeda, V. Kuzmichev, S. Minaev, V. Pershin, B. Sharkov, R. Vengrov, S. Yaramishev, Commissioning of ITEP 27 MHz Heavy Ion RFQ, *Proceedings of the Seventh European Particle Accelerator Conference. EPAC-2000*, Viena, 2000, p.p. 854-856.
6. G.Belyaev, M.Basko, A.Cherkasov, A.Golubev, A.Fertman, I.Roudskoy, S.Savin, B.Sharkov, V.Turtikov, A.Arzumanov, A.Borisenko, I.Gorlachev, S.Lysukhin, D.H.H. Hoffmann, A.Tauschwitz., Measurements of the Coulomb energy loss by fast protons in a plasma target, *Phys. Rev. E*, 1996, V53, N3. 2701-2707

UNDERLYING PHYSICS OF E-MEVVA OPERATION:
Explaining Past Results, Guiding Future Improvements

Ady Hershcovitch[1]
In collaboration with
V.A. Batalin,[2] A.S. Bugaev,[3] N. DeBolt[4], V.I. Gushenets,[3] B.M. Johnson,[1] A.A. Kolomiets,[2] R.P. Kuibeda,[2] T.V. Kulevoy,[2] E.M. Oks,[3] F. Patton,[4] V.I. Pershin,[2] S.V. Petrenko,[2] D.N. Seleznev,[2] N. Rostoker,[4] A. VanDrie,[4] F.J. Wessel,[4] and G.Yu. Yushkov[3]
[1]*Brookhaven National Laboratory, Upton, NY, USA*
[2]*Institute for Theoretical and Experimental Physics, Moscow, RUSSIA*
[3]*High Current Electronics Institute Russian Academy of Sciences, Tomsk, RUSSIA*
[4]*University of California at Irvine, Irvine, California, USA*

Abstract. Recently substantial enhancement of high ion charge states was clearly observed in both the HCEI and ITEP E-MEVVA ion sources. These experimental set-ups have two different methods of measuring the ion charge state distributions. The results can be considered as a proof of the E-MEVVA principle. These results sparked discussions regarding which physics effects are dominant. Basic physics seems straightforward, an ion charge state in E-MEVVA is determined by the number of collisions with fast electrons versus the number of encounters with neutrals and lower charge state ions during an ion dwell time in the drift channel. However, the fluxes of fast electrons, lower charge state ions, and neutrals encountered by an ion may be a consequence of numerous effects. Factors determining neutral fluxes might be poor vacuum conditions, desorption of adsorbed gas by the electron beam directly or indirectly due to stacking (E-beam reflection) and/or instabilities that cause heating and desorption. Flux and energy of the fast electrons is primarily determined by the electron gun output. But significant contributions from electron beam stacking, instabilities, as well as plasma electron heating, are possible. The various contributions are evaluated to account for past results and to guide future progress.

1. Introduction

Metal Vapor Vacuum Arc (MEVVA) ion sources [1] are used to generate high current pulsed ion beams for both fundamental [2] and applied [3] research. The MEVVA is a prolific generator of highly ionized metal plasma from which metallic ions are extracted.

E. Oks and I. Brown (eds.),
Emerging Applications of Vacuum-Arc-Produced Plasma, Ion and Electron Beams, 51–57.

A generic MEVVA [1] consists of a series of electrodes (usually concentric) that are separated by ceramic insulators. The commonly used configuration is a solid electrode of the desired metal, followed by a trigger electrode, an anode, a suppressor, and a three-grid extractor. Triggering of the vacuum arc is accomplished by applying a short high voltage pulse between the trigger electrode and the cathode across an insulating surface. Vacuum arc discharge occurs due to formation of cathode spots, which are micron-sized spots on the cathode surface characterized by extremely high current densities. Small spots on the cathode material are vaporized and ionized, producing a plasma plume, from which ions are extracted. Although a MEVVA plasma is characterized by a high degree of ionization, only low ion charge states are typically extracted. Depending on the cathode material used a conventional MEVVA ion beam has a mean charge state Q of about 2+.

For many applications [2,3] it is highly desirable to enhance the MEVVA ion charge state so that the ion beam energy can be increased without applying higher extraction voltage. Previous efforts demonstrated that the mean ion charge state in vacuum arc plasmas could be increased in a strong magnetic field [4,5], with high arc current [5], or by applying an additional short current "spike" on top of the main arc current [6]. Most previous attempts to obtain higher charge states quickly reached saturation [7] at charge states only 1.5 to 2 times higher than the conventional MEVVA. However, one promising approach is to attempt ion charge state enhancement using an energetic electron beam. Sources like the Electron-Cyclotron Resonance (ECR) and Electron-Beam Ion Source (EBIS) also use energetic electrons to produce high ion charge states, but the ion beam currents are typically orders of magnitude lower than MEVVA. Hence, the purpose of E-MEVVA is to obtain both large ion currents and high charge states.

Over 30 years ago Donets invented the EBIS [8], in which a high-energy, high-density electron beam produced multiple ionization of gaseous ions. Later, Batalin, et al. [9] combined an electron beam, a vacuum arc ion source, and a drift tube into a source called E-MEVVA, which produced encouraging indications of higher charge state production. Then, Hershcovitch, et al. [10] extended this concept using a Z-discharge plasma to generate an internal electron beam. With a gold cathode this Z-MEVVA gave results with some indication [10] of charge states as high as Au^{6+}.

Recently, significant charge state enhancement was report [11,12] in detailed E-MEVVA investigations, which were performed jointly among the Institute for Theoretical and Experimental Physics (ITEP), Moscow, Russia, the High Current Electronics Institute (HCEI), Tomsk, Russia, and Brookhaven National Laboratory (BNL), USA. The experiments were performed in Moscow and Tomsk with nearly the same design of ion sources. Substantially higher ion charge states were observed clearly in both experimental set-ups with two different methods of measuring the ion charge state distributions.

In this paper the underlying E-MEVVA physics is reviewed. Old results are interpreted, future improvements are proposed.

2. Physics of Charge-State Enhancement and Reduction

When electron-impact dominates ionization like in E-MEVVA, three ingredients are needed: (1) high $J\tau$, which is the product of electron current density J and electron-ion interaction time τ, and (2) high E, which is the effective electron "beam" energy. Donets [13] is credited with illustrating that the maximum charge state achievable for any element can be predicted on a plot of $j\tau$ versus E. Additionally, (3) prevention of charge reduction, which is dominated by exchange with lower charge state ions and neutrals.

When stepwise ionization, by electrons with density n_e and velocity v_e, is the dominant stripping process, the equation describing the rate of change in the number N_q of ions in a charge state q is

$$dN_q/dt = - N_q\, n_e \sigma_{q\rightarrow q+1} v_e + N_{q-1}\, n_e \sigma_{q-1\rightarrow q} v_e \qquad (1)$$

where σ is the cross section for ionization of ground-state ions. A reasonably good expression for σ is Lötz's semi-empirical ionization formula [14],

$$\sigma_{q\rightarrow q+1} = 4.5\text{x}10^{14}\, \Sigma\, (n_j/EI_j)\, \ln(E/I_j) \quad (\text{cm}^2) \qquad (2)$$

where n_j is the number of electrons in subshell j, I_j is the ionization energy of subshell j in eV and E is the electron incident energy in eV. In the absence of any other processes Eq. 1 can be integrated to yield an expression describing the time evolution of the number N_q of ions in a charge state q as a function of $J\tau$. The "Donets plot" [13] is obtained by plotting the minimal E required to reach a charge state versus the $J\tau$. Unlike in EBIS, charge exchange is a very important contributor in plasma heavy-ion sources like the MEVVA and E-MEVVA, because the high charge-state ions interact with newly formed plasma ions and background atoms. To include the effect of charge exchange requires adding to Eq. 1 an additional term

$$dN_q/dt = -N_q\, n_e \sigma_{q\rightarrow q+1} v_e + N_{q-1}\, n_e \sigma_{q-1\rightarrow q} v_e \;-\; \Sigma\, N_q\, n_i\, \sigma_{cq\rightarrow q-1}\, v_i \qquad (3)$$

where, $\sigma_{cq\rightarrow q-1}$ is the single electron-capture cross section by charge exchange with ions in the discharge with charge state less than q. These ions have a variety of charge states, n_i and v_i are density and relative velocity (to ions with charge q) in charge state $i<q$. For $\sigma_{cq\rightarrow q-1}$ there is a simple semi-empirical formula that describes the dependence of this cross section on q and on v_i [15]

$$\sigma(q,v_i) \propto q^a/v_i^m \qquad (4)$$

where the parameters a and m are to be determined from either experimental or theoretical work. In studies with MeV projectiles [15], the value of a was estimated and measured in the range of 2-3.7, while m was 3-4. Our interest is in a much lower (keV) energy range where the value of a may be even larger than 3.7 [16]. Multi-electron capture is rather significant for highly-charged ions, as observed in Kr^{+18} - Ar collisions [17]. A more realistic version of Eq. 4 would require inclusion of multi-electron capture; however only limited data is available. Nevertheless, Eqs. 2-4 indicate that in sources with continuous plasma formation very high charge states cannot be attained in large quantities. The stripping cross section decreases with increase in ionization energy (i.e., charge state), while the electron-capture cross section increases with charge state. Plasma formation rates in heavy ion sources (in which plasma is continuously formed) are usually large enough to result in a significant density of low charge state ions, which in turn suppress generation of high charge state ions. In vacuum arcs with currents of a few hundred Amperes, e.g., typical cathode erosion rate is about 30 μg/Coulomb[18] resulting in an ion current that is roughly 10% of the total arc current [19].

Equations 1 and 3 are based on stepwise ionization of ground state ions. However, charge state formation rates higher by a factor of 2.5 have been observed in Z-pinches [20]. A number of additional contributions may lead to the higher rates, e.g., ionization of excited ions with a cross section larger than Eq. 2; and, excitation - autoionization (Auger) processes. In most plasma heavy ion sources like the EBIS, ECR, PIG, and MEVVA, excited ions decay before collisions leading to ionizations occur. At higher charge states in a typical EBIS, the time interval between successive ionizations is at least a number of milli-seconds, i.e., orders of magnitude longer than the decay time of most excited ions, whereas the whole ionization process in an E-MEVVA lasts for microseconds. The same arguments can be extended to with an intense electron beam. Therefore, Eq. 3 must be modified for such intense devices to include autoionization, ionization of excited ions. Including those contributions yields,

$$dN_q/dt = \Sigma^* (-n_q n_e \varsigma_{q\rightarrow q+1} v_e + \mathbf{n}_{q-1} n_e \varsigma_{q-1\rightarrow q} v_e) - N_q n_i s_{iq\rightarrow q+1} v \qquad (5)$$

where Σ^* refers to summation over all ion states (ground and excited) $\mathbf{n}_q$ is the density of each state; the total ionization cross section by electron impact $\varsigma = \sigma^* + \sigma^{s+a}$ in which, σ^* in the ionization cross section of excited ion (for which there is no analytical expression and very little data) and σ^{s+a} is the total impact ionization of ground state ions by electron stripping as well as autoionization [a semi-empirical formula for σ^{s+a} can be found in [21]. These terms account for ionization by background ions. A procedure for computing σ_i can be found in [22].

3. Discussion

Equation 3 and the ensuing discussion clearly indicate that, in discharges with continuous formation of neutrals and low charge state ions, very high charge state heavy ions cannot be attained in significant quantities. To illustrate this charge-exchange limitation consider the charge changing cross sections [15] of iodine ions passing through a hydrogen target, which for 5 MeV I^{+7} are: 18.5 Å^2 for electron capture (i.e., charge exchange resulting in I^{+6}), and 0.045 Å^2 for electron loss (i.e., ionization resulting in I^{+8}) respectively. As predicted by equations 3 and 4, the data shows that the ratio (of over 400) between these processes (cross sections) is rather unfavorable for high charge state formation. Since the I^{+7} energy is much larger than the hydrogen binding energy, the electron loss cross section is equivalent to ionization by free electrons with an equal relative velocity (as would be the case in an ion source). However, in any conceivable (useful) ion source, the ion energy spread would not exceed a few KeV. Hence, based on equation 5, the electron capture cross section in an ion source would be much higher than that measured in [15]. Furthermore, the data and Eq. 3 indicate a worsening of cross section (charge-exchange/ ionization) ratios with increase in charge state, e.g., the ratio which is (3.54 $\AA^2$)/(3 $\AA^2$) = 1.18 for I^{+2} grows to 411 for I^{+7}.

A simple model [23,24] assumes that the ion charge state distribution in an E-MEVVA (and in some other ion sources), is determined by the balance of electron stripping rate versus neutralization by charge exchange with neutrals and lower charge state ions. The relative fraction of higher charge state ions can be enhanced by raising the intensity of the electron beam in the drift region, and by preventing "fresh plasma" formation during stripping, thus reducing the undesirable effects of charge exchange. To enhance E-MEVVA ion charge states the electron beam currents in the drift tube were raised [11,12] from 1 A to 40 A. To curtail charge state reduction by charge exchange the vacuum systems were improved and the E-MEVVA electron beam pulse was made longer than the MEVVA pulse. If no fresh plasma is generated during most of the electron beam pulse, the unfavorable charge exchange is greatly reduced. Electron stripping to higher charge states becomes the dominant process.

The basic motivations for the improvements above are clear and difficult to dispute; however, implementing these changes resulted in ion charge state distributions that either failed to show any charge state enhancement or showed a mild reduction in high charge state fractions. The culprit was gas generation by the electron beam, which compounded the problem of gas generation by the MEVVA arc. When the electron beam is fired, the impurity ion population increases dramatically due to electrons striking the drift tube walls. The breakthrough [11] came when the MEVVA arc was lowered and the electron beam pulse length was shortened to reduce gas generation.

Surprisingly, the recent E-MEVVA results [11,12] are consistent with $J\tau$ predictions. That is, successive single (stepwise) ionization accounts for the observations. Given the relatively poor vacuum condition during the electron beam pulse, unfavorable charge exchange conditions are likely. The apparent agreement with "$J\tau$ scaling" is most likely the result of multiple ionizations compensating for destructive charge exchange. For high charge states multiple ionization by single-electron impact is greatly reduced. Therefore, reducing gas and impurity ion density is imperative. $J\tau$ must be increased to attain higher charge states.

Strong evidence also exists for electron beam stacking, which has the convoluted contributions of enhancing $J\tau$, while sputtering impurity off the walls. It leads to instabilities, which heat the plasma and increase impurity concentrations. The results are also consistent with an alternative interpretation by A. Anders [25], who suggests that a small increase in plasma electron temperature (resulting from electron beam heating) can significantly increase the population of energetic electrons, and hence, ion charge states.

4. Conclusion

Although E-MEVVA has clearly shown to produce substantially higher charge-state ions than a conventional MEVVA, it may be possible to further optimize the source and to extract even higher charge state ions after the electron beam pulse. Possibilities for future enhancement include (a) increasing the electron beam current and density, and (b) further reducing the negative effect of residual gas impurities. Increasing the electron beam current and density is a straightforward concept. However, the electron gun would have to be completely gasless, unlike the present E-MEVVA electron guns. To prevent formation of impurities, the electron beam, after passing through the drift tube, would be guided into an external beam dump. The beam dump must face away from the ion beam axis to prevent gaseous impurities from streaming into regions where they can interact with the ions. To generate very high charge states, a merging beam approach would be needed.

ACKNOWLEDGEMENTS

The authors gratefully acknowledge I.G. Brown and A. Anders (LBNL, Berkeley) and P. Spadtke (GSI, Darmstadt) for fruitful discussions. This work is supported by Research contract between BNL and ITEP with HCEI under the IPP Thrust-1 program, by Russian Foundation of Basic Research under Grant No. 99-02-18163 and by Grant of Russian Ministry of Education for Fundamental Research.

References

1. I.G. Brown, Rev. Sci. Instrum. **65**, 3061 (1994).
2. H. Reich, P. Spadtke, and E.M. Oks, Rev. Sci. Instrum. **71**, 707 (2000).
3. N.V. Gavrilov and E.M Oks, Nucl. Instrum. Meth. in Phys. Res. **A 439**, 31 (2000).
4. E.M. Oks, I.G. Brown, M.R. Dickinson, R.A. MacGill, P. Spadtke, H. Emig, and B. Wolf, Appl. Phys. Letters. **67**, 200 (1995).
5. E.M. Oks, A. Anders, I.G. Brown, M.R. Dickinson, and R.A. MacGill, IEEE Trans. on Plasma Science. **24**, 1174 (1996).
6. A.S. Bugaev, E.M. Oks, G.Yu. Yushkov, A. Anders, and I.G. Brown, Rev. Sci. Instrum. **71,** 701 (2000).
7. A. Anders, G. Yu. Yushkov, E.M. Oks, A.G. Nikolaev, and I.G. Brown, Rev. Sci. Instrum. **69,** 1332 (1998).
8. E. Donets, Rev. Sci. Instrum. **69,** 614 (1998).
9. V. A. Batalin, Y. Volkov, T.V. Kulevoy, and S.V. Petrenko, *Proceedings of the 4th European Accelerator Conference* (EPAC-94, London, England) 1560 (1994); *Proceedings of the 17th International LINAC Conference* (LINAC-94, KEK, Ibaraki, Japan) 390-392 (1994).
10. A. Hershcovitch, B.M. Johnson, F. Liu, A. Anders, and I.G. Brown, Rev. Sci. Instrum. **69,** 798 (1998).
11. A.S. Bugaev, V.I Gushenets, E.M. Oks, G.Yu Yushkov, T.V. Kulevoy, A. Hershcovitch and B.M Johnson, Appl. Phys. Lett., **79,** 919 (2001).
12. V.A Batalin, A.S. Bugaev, V.I. Gushenets, A. Hershcovitch, B. M. Johnson, A.A. Kolomiets, R.P. Kuibeda, T.V. Kulevoy, E.M. Oks[2], V.I. Pershin, S.V. Petrenko, D.N., Seleznev, and G.Yu. Yushkov, Rev. Sci. Instrum. **73**, 702 (2002).

13. E.D. Donets, in Proceedings of Fifth All-Union Conference on Charged Particle Accelerators, Nuaka, Moscow (unpublished), **1**, 346 (1977).
14. W. Lötz, Z. Physik **216**, 241(1968); and, **220**, 466 (1969).
15. H. D. Betz and A.B. Wittkower, Phys. Rev. A **6**, 1485 (1972).
16. A. Hershcovitch, Rev. Sci. Instrum. **65**, 1075 (1994).
17. S. Martin, A. Devis, Y. Querdane, A. Salmoun, A. El Motassadeq, J. Desequelles, M. Druetta, D. Church, and T. Lang, Phys. Rev. Lett.**64**, 2633 (1990).
18. *Vacuum Arcs Science and Technology*, R.L. Boxman, P.J. Martin, and D.M. Sanders, Editors, Noys, New York (1995).
19. I. Brown, H. Shiraishi, IEEE Transactions of Plasma Science **18**, 170 (1990).
20. C. Stöckl, H. Wetzler, W. Seelig, J. Jacoby, P. Spiller, and D. Hoffmann, GSI Sci. Report 1994 **94-1**, 169 (March 1995).
21. Burgess and Chidichimo, Mon. Not. R. Astr. Soc. **203**, 1269 (1983).
22. McGuire and Richard, Phys. Rev. A **3**, 1374 (1973).
23. A. Hershcovitch, B. Johnson, F. Patton, N. Rostoker, A. VanDrie, and F. Wessel, *Electron Beams and Z-Pinches As Plasma Strippers and Lenses for Low Energy Heavy Ions*, BNL Internal Report: AGS/AD/Technical Note No. 484 (1999).
24. A. Hershcovitch, B. Johnson, F. Patton, N. Rostoker, A. VanDrie, and F. Wessel, Rev. Sci. Instrum. **73**, 744 (2002).
25. A. Anders private communication 2001.

TECHNICAL DESIGN OF THE MEVVA ION SOURCE AT GSI AND RESULTS OF A LONG URANIUM BEAM TIME PERIOD

F. HEYMACH, M. GALONSKA, R. HOLLINGER, K. D. LEIBLE,
P. SPÄDTKE, M. STORK

Gesellschaft für Schwerionenforschung mbH,
Planckstraße 1, D-64291 Darmstadt, Germany

and

E. Oks

High Current Electronics Institute,
Russian Academy of Sciences, Tomks, Russia

Abstract. The design ion for the linac at GSI is $^{238}U^{4+}$. The goal is to achieve a high current ion beam to fill the synchrotron up to the space charge limit with good reproducibility and high availability. An overview of the MEVVA ion source with the most important operational data and technical details for uranium operation will be given. Further improvements of the source have been carried out: Meshes of stainless steel are used in the plasma drift section instead of grids used in earlier experiments (different material, place, shape). With these meshes it was possible to increase the discharge current to get better spectra, better pulse stability and lower noise. To increase the ion current and the reliability we have modified the electrode shaping of the extraction system and replaced its tungsten copper alloy by molybdenum. A new glue technique for the trigger ring has improved the trigger reliability. Some results of the 300 hours uranium beam time in December 2001 at the UNILAC will be presented. We will give an overview about the beam transport along the low energy beam.

1. Introduction

The challenge of the MEVVA for the application at the GSI facility was the pulse-to-pulse stability and the noise within the macro-pulse [1–7]. A stable ion beam is essential among other things for accelerator transmission optimization. With our improvements we could use the ion source for operation at the accelerator. The ion source was optimized for $^{238}U^{4+}$ and proved to be reliable over a long beam time period. The beam transport from the source to the beam line was investigated in 2001 with a MUCIS [8] ion source and an argon beam [9]. Both ion sources, the MEVVA and the MUCIS, use

E. Oks and I. Brown (eds.),
Emerging Applications of Vacuum-Arc-Produced Plasma, Ion and Electron Beams, 59–65.

the same extraction system, delivering nearly the same ion beam emittance [10]. With several improvements it was possible to increase the brilliance to ensure that a sufficient beam current could be transported from the terminal to the low energy beam line. Further on we made some improvements concerning the extraction system, meshes, and the cathodes. In this article we report mainly on the experimental setup and technical solutions. Theoretical aspects are covered in [11].

2. The source

For the production of high current metal ion beams a (MEVVA) metal vapor vacuum arc ion source is used [12]. The source is operated in a pulsed mode of 0.5–1 Hz, with typical 1 ms pulse length. Figure 1 shows the MEVVA ion source.

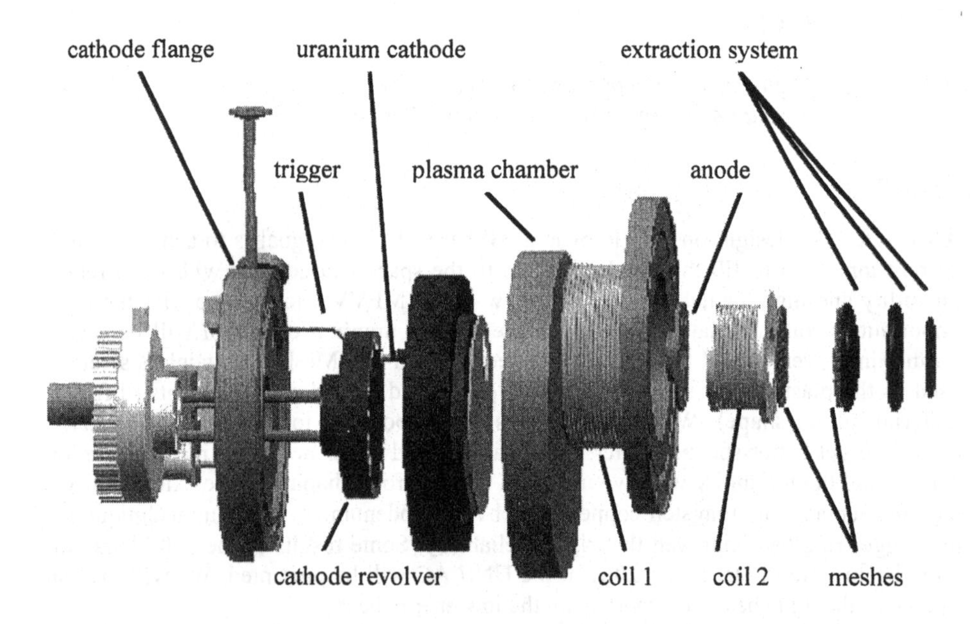

Figure 1. Exploded view of the MEVVA ion source at GSI.

On the left hand side is the cathode flange with a movable cathode revolver which can carry 17 cathodes. Each cathode has a diameter of 5.7 mm. The distance between the cathode and the anode is 7 mm. The aperture of the anode is 13 mm. Coil 1 is a dc solenoid to confine the plasma. Coil 2 is pulsed, up to 100 A, typical 40 A, to enhance higher charge states. Between the pulsed coil and the accel-decel extraction system are two meshes one connected to coil 2 and one to the extraction. At the moment we are using stainless steel meshes in a frame, mesh width 0.335 mm and wire diameter 0.14 mm. The lifetime of the meshes under our operating conditions is of the order of one week. The extraction system consists of 13 x Ø3 mm aperture geometry. Operational data are shown in table 1.

TABLE 1. Operational data and source parameters (see figure 2 for I_{UL3DC} and I_{UL4DT4})

timing	1 Hz, 0.5–1ms
arc current	800A
arc power	40 kW
emission current density	60 mA/cm^2
$^{238}U^{3+}$	20 %
$^{238}U^{4+}$	67 %
$^{238}U^{5+}$	13 %
I_{UL3DC}	55 mA (30kV) full beam
I_{UL4DT4}	25 mA
noise U^{4+}	±10%
cathode life time	10 hours

3. The Facility

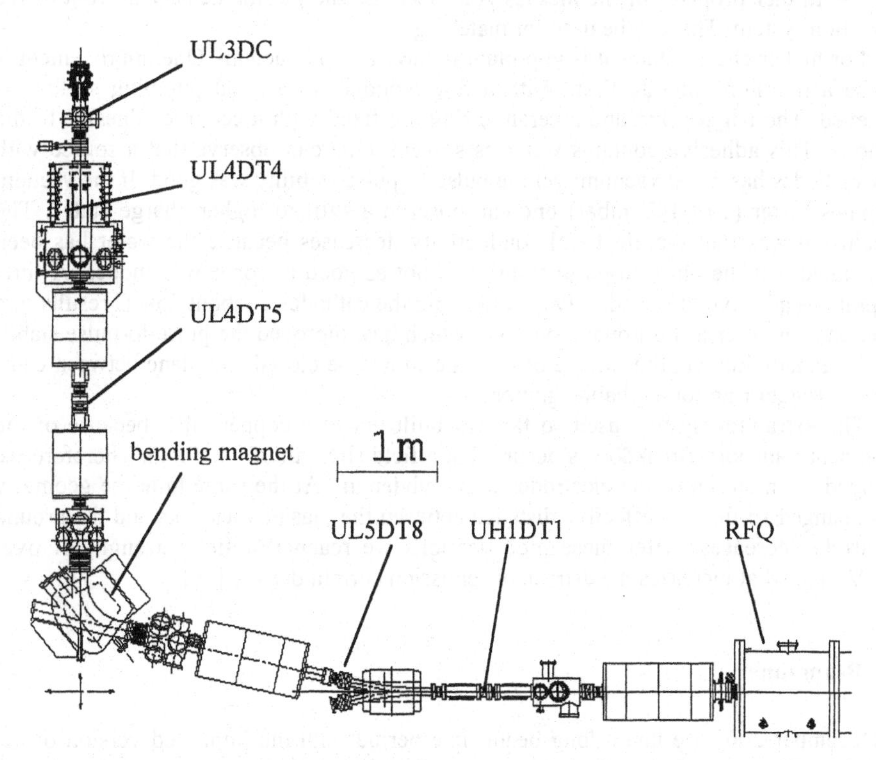

Figure 2. The low energy beam line (LEBL) from the source to the RFQ with beam transformers (DT) and Faraday cup (DC).

Figure 2 shows the main elements of the low energy beam line. The source is on a 130 kV terminal. The ions are post-accelerated DC. The mass and charge separation takes place in the LEBL (bending magnet). Our goal is to fill the RFQ at the front end of the UNILAC to its space charge limit of 0.25 A/ζ [mA]. The desired energy of the RFQ is 2.2 keV A/ζ. The acceptance of the RFQ is 138 π mm mrad [13, 14]. The acceleration system consists of a single gap with integrated screening electrode. This gap is 0.5 m behind the ion source extractor between is a drift section. For $^{238}U^{4+}$ the required voltage is 130.9 kV, separated into 35 kV extraction voltage and 95.9 kV acceleration voltage.

4. Observations, properties and improvements

Metal meshes (stainless steel, nickel, molybdenum and tungsten) were used in the past [5, 15] to influence the behavior of the source. Some of them are too weak and the plasma plume destroyed them. If the mesh apertures are too big there is only a small influence on the noise and if the mesh apertures are too small then the transmission is too bad and they are engorged by macro particles very fast. The position of the meshes plays an important role. For our requirements the meshes are fixed directly to the anode for lower charge states e.g. Ni^{1+} or in the extraction region for higher charge states e.g. U^{4+}. A further property of the meshes is to decrease the plasma density in front of the extraction system. This can be used for matching.

For higher charge states it is important to have a good vacuum. One improvement is to use a titanium cathode first. With a few monolayers a good pumping surface is achieved. The trigger ring and a ceramic ring are fixed with a ceramic adhesive to the cathode. This adhesive contains water as solvent. One can observe that a source with new cathodes has a bad vacuum; yet the pulse-to-pulse stability was good. If the vacuum becomes better (10^{-6}–10^{-7} mbar) one can observe a shift to higher charge states. The conclusion was that the electrical conductivity decreases because the water has been evaporated, yet the pulse-to-pulse stability is not as good as for new cathodes. A little graphite as additive in the adhesive, and baking the cathodes at about 250°C results in a good vacuum after a short operation time, which has improved the pulse-to-pulse stability. As already known, the surface of the glue should be closed and plane between cathode and trigger ring for a reliable ignition.

The extraction system used so far was built out of a copper alloy because of the good heat transport. Breakdowns occurred at a field strength of 8 kV/mm. Therefore we changed the material of the electrodes to molybdenum. At the same time the geometry was changed so that the effective distance between the plasma meniscus and the ground electrode decreases. After these modifications we reached a field strength of over 11 kV/mm, what increases the extractable emission current density [16].

5. Beam time

In December 2001 we had a long beam time period with the improved version of the MEVVA. Figures 3–5 show the online measurement of the ion current (each pulse recorded) along the low energy beam line.

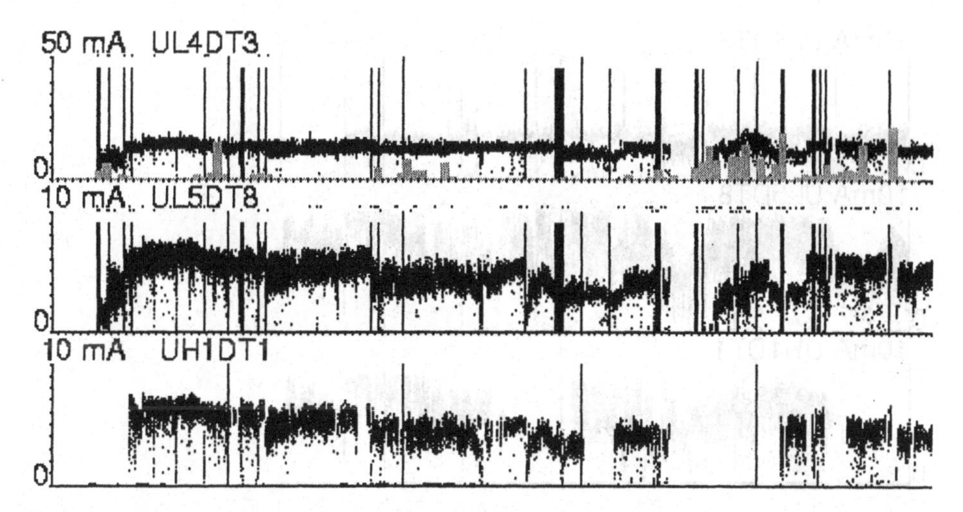

Figure 3. The first five days of the uranium beam time in December 2001. Data of three beam transformers (BT) are diagrammed see (figure 2). At the bottom is the BT UH1DT1 in front of the RFQ. The vertical lines in this row are the days. The other vertical lines are cathode changes or cathode tests. The gray bars at the top row UL4DT3 (total ion beam current) symbolize the number of operator interactions. The other two BTs show the U^{4+} fractions of the ion beam after the bending magnet. An interesting remark: in the bottom row of figures 3–5 one can see a second beam (Ca) from a PIG ion source. This is the thicker line direct on the time axis.

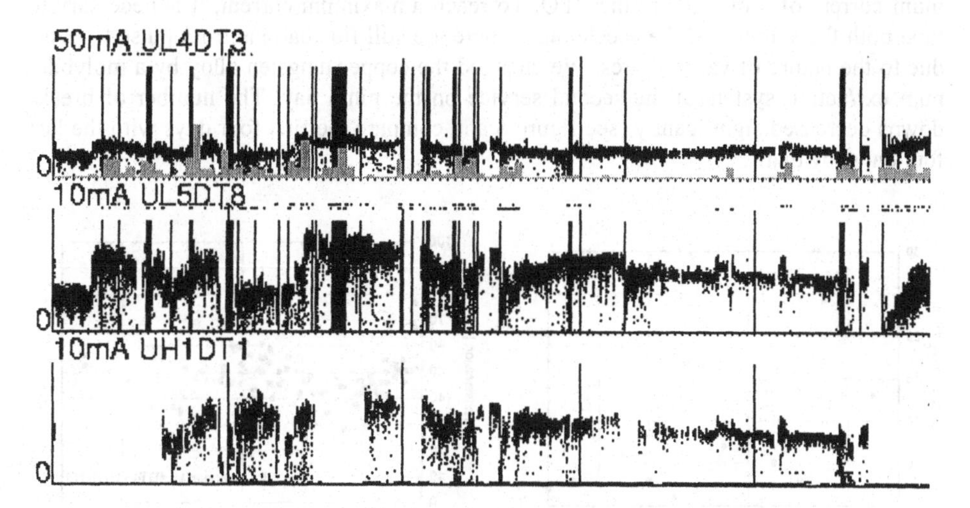

Figure 4. The next five days of the uranium beam time in December 2001.

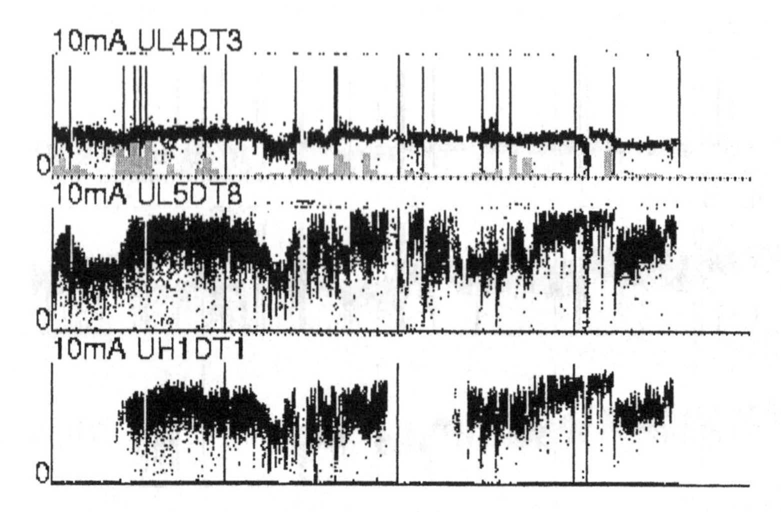

Figure 5. The last days of the uranium beam time in December 2001.

Servicing the source was necessary on the fourth and on the ninth day. A new source needs six hours to deliver a sufficient current after each service (outgassing cathode, pulsed coil) and up to two hours after a cathode change. Our goal was to deliver a minimum current of 5 mA U^{4+} to the RFQ. To reach a maximum current, it is necessary to tune both the source and the accelerator. There are still fluctuations from pulse to pulse due to the nature of vacuum arcs. We changed the copper tungsten alloy by a molybdenum extraction system at the second service on the ninth day. The number of breakdowns decreased significantly (see figure 6 and compare the first four days with the last four days) by that.

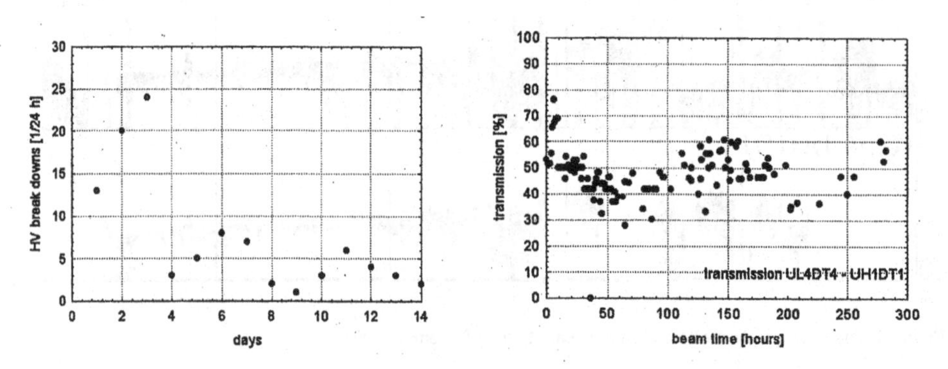

Figure 6. left: HV breakdowns of the extraction voltage at 30kV for the first four days, 25 kV for the next days, up to 35 kV for the last four days; right: Transmission through the LEBL

6. Conclusion

The beam time was a success for the MEVVA ion source. It has been demonstrated that an application at our facility is possible. For the future we should work on the transmission and decrease the losses. The new material for the extraction system decreases the electrical breakdowns enormously. Some effects, e.g. from the pulsed coil and the anode material should be investigated. Further efforts should be undertaken in order to make the service more easy and safer. We think we can optimize the source after further investigations resulting in future improvements.

References:

1. B. H. Wolf, H. Emig, D. Rück, P. Spädtke: Investigation of MEVVA ion source for metal ion injection into accelerators at GSI, *Rev. Sci. Instrum.* (1994), **Vol.~65 no.~10**, pp.3091–3098.
2. P. Spädtke: U^{4+} MEVVA source & outlook for HIF sources, *Nucl. Instr. and Meth. in Phys. Res.* (2001), **A 464**, pp. 388–394.
3. J. E. Galvin, I. G. Brown, R. E. MacGill: Charge state distribution studies of the metal vapor vacuum arc ion source, *Rev. Sci. Instrum.* (1990), **Vol. 61 no. 1**, pp. 583–585.
4. F. J. Paoloni, I. G Brown: Some observations of the effect of magnetic field and arc current on the vacuum arc ion charge state distribution, *Rev. Sci. Instrum.* (1995), **Vol. 66 no. 7**, pp. 3855–3858.
5. A. Anders, R. Hollinger: Reducing ion-beam noise of vacuum arc ion sources, *Rev. Sci. Instrum.* (2002), **Vol. 73**, pp. 732–734.
6. E. Oks, A. Anders, G. Brown et al.: Ion charge state distributions in high current vacuum arc plasmas in a magnetic Field, *IEEE Transaction on Plasma science* (1996), **Vol. 24 no. 3**, pp. 1174–1183.
7. A. Anders, G. Yushkov, E. Oks, A. Nikolaev, I. Brown: Ion charge state distributions of pulsed vacuum arc plasmas in strong magnetic fields, *Rev. Sci. Instrum.* (1998), **Vol. 69**, pp. 1332–1335.
8. R. Keller: Multicharged ion production with MUCIS, *GSI Annual Report*, (1987), p. 360
9. R. Hollinger, L. Dahl, F. Heymach, P. Spädtke: Transversale Emittanzmessung am Injektor Nord, *GSI Report 2001-02 Juli* (2001).
10. H. Reich, F. Heymach, P. Spädtke: Commissioning of the high current ion sources at the new GSI injector (HSI), *Proc. LINAC 2000*, **Vol. 1**, pp. 238–240.
11. M. Galonska these proceedings.
12. S. Anders, A. Anders, I. Brown: Vacuum arc ion sources: Some vacuum arc basics and recent results, *Rev. Sci. Instrum.* (1994), **Vol. 65 no.4**, pp. 1253–1258.
13. W. Barth, L. Dahl, J. Klabunde, C. Mühle, P. Spädtke: High-intensity low energy beam transport design studies for the new injektor linac of the UILAC, *Proc. LINAC96*, **Vol. 1**, pp. 134–136.
14. P. Spädtke, H. Emig, J. Klabunde, D. M. Rück, K. Tinschert: Acceleration of high-current ion beams, *Rev. Sci. Instrum.* (1996), **Vol. 67 no. 3**, pp. 1146–1148.
15. E. Oks, P. Spädtke, H. Emig, B. H. Wolf: Ion beam noise reduction method for the MEVVA ion source, *Rev. Sci. Instrum.* (1994), **Vol. 65 no. 10**, pp. 3109–3112.
16. C. D. Child: *Phys. Rev.* (1911) **(Ser. 1) 32**, 492

SIMULATION OF THE EXTRACTION FROM A MEVVA ION SOURCE

PETER SPÄDTKE
GSI Darmstadt
Planckstr. 1
64291 Darmstadt
GERMANY

Abstract. The specific ingredients to simulate the extraction from a MEVVA ion source will be described. These parameters are: the particle density in front of the extraction system, the energy of the electrons inside the plasma, the charge state distribution of the extracted beam, and the initial energy of the ions, geometry, and applied potentials. The effect of partial space charge compensation will be discussed.

1. Introduction

To achieve a higher particle number for the synchrotron SIS the high current injector HSI has been designed and built [1] and was taken into regular operation in 1999 [2]. For elements available in gaseous form we are using a cusp ion source [3], whereas for metallic elements a vacuum arc ion source [4] is used, but an identical extraction system is applied to both sources. For the commissioning of the accelerator of RFQ and IH type a singly charged argon ion beam has been used, the design ion however is uranium with charge state 4. The ion source has to deliver to the RFQ a current of 0.25 mA $\times$ A/ζ with an energy of 2.2 keV $\times$ A, where A is the atomic mass, and ζ is the charge state. The available transversal acceptance of the RFQ is 138 π mm mrad. The source has to be operated in a pulsed mode with up to 16 pulses per second and a pulse length of up to 1 ms. Typical operating conditions, however, were 1 Hz and 0.5 ms pulse length.

With our extraction system more than 100 mA total uranium ion current could be extracted and up to 30 mA could be analyzed in U^{4+} at the test bench [5]. Because of the design value of 16.5 mA U^{4+} for the theoretical space charge limit of the RFQ we felt well prepared to provide this beam for routine operation. It turned out however, that only a fraction of 3...5 mA could be delivered in earlier experiments to the entrance of the RFQ. The differences between these two experiments are the different lengths of the transport channel, the additional acceleration gap, and slightly different magnetic lenses. The beam line is shown in Fig. 1.

Whereas some losses along the beam line in case of argon beam could be tolerated easily, this is not acceptable for ions required in a higher charge state

E. Oks and I. Brown (eds.),
Emerging Applications of Vacuum-Arc-Produced Plasma, Ion and Electron Beams, 67–77.

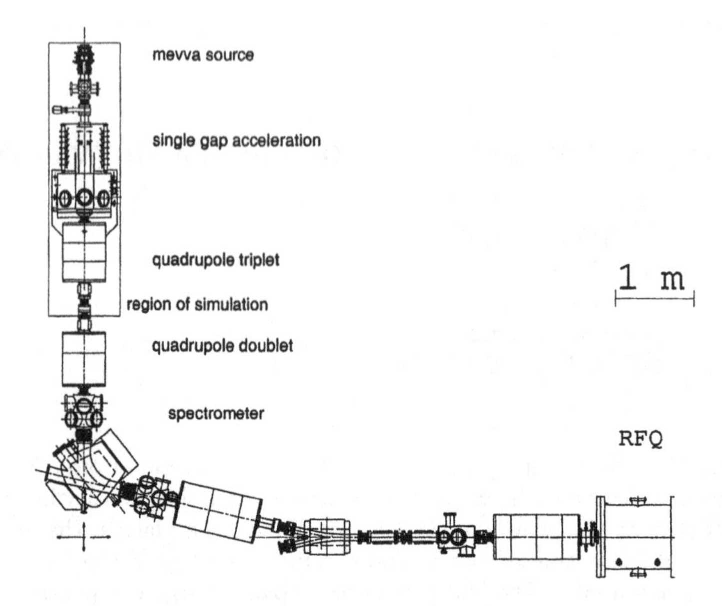

Figure 1: Beam line with Mevva source, acceleration gap, and spectrometer. The region of computation is indicated by a box.

because all other unwanted charge states have to be transported as well and the total load of the highly regulated HV power supply is limited. More important, the maximum ion current which can be handled in the acceleration gap is limited due to optical reason. This is why ongoing simulations are used to improve the state of beam generation, acceleration, and transport.

The strategy for this simulation with the 3D program KOBRA3-INP [6] was to use all available experimental data without any fitting or correction. Then unknown parameters have been varied until the simulation corresponds with the experimental results. These results have been obtained during a two weeks experiment at the UNILAC in December, 2001 [7]. The investigated part of the beam line is divided into four sections because the different geometrical problems require a different resolution which cannot be handled with reasonable memory requirements for the PC, see Tab. 2.

1. extraction
2. drift from extraction to post acceleration
3. post acceleration gap
4. drift and transport through the first quadrupole triplet

The coordinates of all trajectories in a given section are saved at a certain location to be used as starting conditions in the next section. The fraction of transported U^{4+} behind the first quadrupole triplet is very sensitive to modifications of starting conditions within the source. If a solution can be found which behaves as the ion beam in experiment, modifications of the existing beam line to improve are more easy and reliable to predict.

2. Experimental Setup

2.1 Extraction

The extraction system is a conventionally accel-decel extraction system with 13 holes each 3 mm in diameter. The distances between the extraction electrodes are 3 mm, and 1 mm, the applied voltages are 30 kV, and -1 kV. The extraction system is shown in Fig. 2.

A plasma potential of 50 Volt is assumed in the simulation. The charge state distribution has been measured under the actual operating conditions of the source. The numbers are given in Tab. 1.

Table 1: charge state distribution

U^{2+}	U^{3+}	U^{4+}	U^{5+}	U^{6+}
0%	23%	67%	10%	0%

Because the total extracted beam current is an input parameter the beam profile and the current which can be injected into the acceleration gap are good indicators for plausible assumptions for the simulation.

After several days of operation of the source the grid between anode and extraction electrode becomes engorged by macro particles coming from the cathode. This diameter of that area which we assume to be the plasma diameter is much smaller than the diameter of the extraction system. An inhomogeneous plasma density distribution before extraction can be assumed therefore. This was taken into consideration for the generation of starting coordinates in the simulation. The outer six holes are operated with lower current density resulting in a bad matching (because of the very low dense plasma density). This is shown in Fig. 3, which is a projection of the geometry and the trajectories into one plane.

2.2. Drift

The main assumption for this section is the presence of space charge compensation. The origin of the compensating electrons are collisions of primary ions with residual gas atoms, and sputtering of the primary beam on electrodes and surrounding surfaces. These electrons will be trapped by the potential of the uncompensated beam, lowering this potential until it will be as low as the electron temperature.

An assumption of 95% of space charge compensation or higher is necessary to achieve the experimental observed transmission. This compensation is build up within less then 100 μs. The pressure in our beam line is about 10^{-7} mbar.

2.3. Post acceleration

The beam with all different charge states enters the post acceleration gap with the initial energy of 30 kV $\times \zeta$. The initially space charge compensated beam becomes non neutralized while the beam is under the influence of electric fields, as for example in the acceleration section. A plasma boundary similar to the plasma boundary in the extraction system of the ion source is created, see Fig. 7. The negatively biased electrode behind the acceleration gap preserves the space charge compensation built up again on ground potential. A potential of -5 kV was sufficient in all experiments.

It should be pointed out that an optimum exists for the ratio of the extraction voltage and the residual voltage necessary within the acceleration gap to achieve the injection energy into the RFQ. This ratio can be influenced in a certain range by our moveable gap geometry. In our case the extraction voltage was 30 kV, and the acceleration voltage was 100.9 kV with a gap length of 50 mm.

The coordinates of the fully accelerated ion beam trajectories are transferred to the next section, where the first lens, a magnetic quadrupole triplet, has to accept the beam. Because of the magnetic force this is the first location where the mixture of charge states are treated differently according their A/ζ ratio.

2.4. Drift and magnetic quadrupole triplet

The aim of this lens is to produce a parallel beam with large diameter which is necessary for the following quadrupole doublet and a 77°-dipole magnet. The gradients of the triplet are taken exactly from experiment. From all experiments we know, that the space charge compensation of the ion beam is higher than 95 % after less than 100 μs as it is in the first drift section.

3. Simulation

Table 2: Numerical resolution and memory requirement

	section	physical size [cm]	number of nodes	memory
I	extraction	$2 \times 3 \times 3$	$201 \times 151 \times 151$	120 MB
II	drift	$50 \times 12 \times 12$	$201 \times 121 \times 121$	80 MB
III	acceleration	$20 \times 12 \times 12$	$201 \times 121 \times 121$	80 MB
IV	quadrupole	$200 \times 12 \times 12$	$201 \times 121 \times 121$	110 MB

The simulation of the extraction of the uranium beam with the measured charge state distribution and the assumed density distribution is shown in Fig. 3, the emittance after extraction is shown in Fig. 4. Whereas the extraction holes close to the center have beam lets with matched current density, the beam lets from the outer extraction holes are over focused due to the low current density.

To investigate the question what the starting velocity of the ions is under experimental conditions, the full beam line has been simulated with different starting energies: 10 eV, 100 eV, and 160 eV, and the case of a common velocity of $160\,\mathrm{eV}/\zeta$ which is because of the explosive generation of the ions [8] has been investigated as well. The amount of U^{4+} after the triplet has been used as indicator. The numbers are summarized in Tab. 3.

Table 3: Influence of different starting conditions of the ions due to the explosive nature of the discharge on the transported current.

starting energy (eV)	current (mA) extracted	accelerated	end of beam line
10	71.3	18.4	17.2
100	56.2	24.7	16.5
160			8.8
100/q	56.1	12.4	11.1

The number of selected starting conditions is still not sufficient to give a final answer, but a strong influence can be observed. Further simulations are necessary.

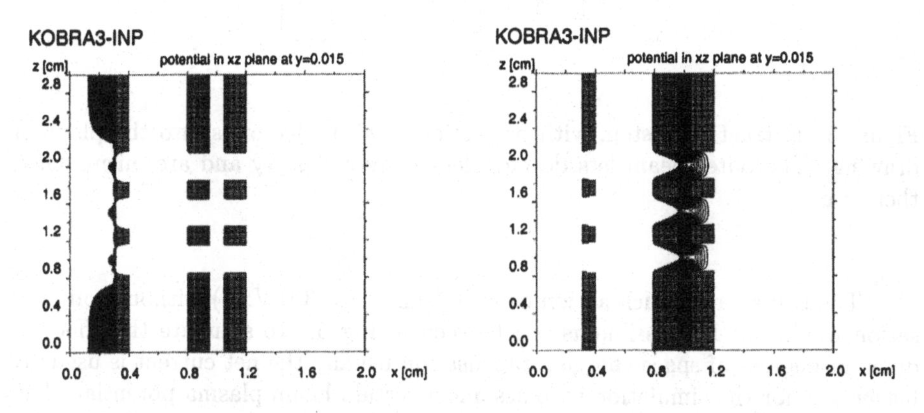

Figure 2: Left: extraction system with plasma boundary. Right: extraction system with negative screening potential.

The plasma boundary is the most important optical element for the extraction. The shape of this boundary depends on the particle density and the velocity distribution of ions and electrons and the applied electric field. In this simulation only cold electrons of 5 eV are assumed. Electrons of higher energies from 10 to several 100 eV would influence this shape considerably. This energy distribution should be measured, or estimated theoretically.

The negative screening potential indicates, that only a few Volts negative potential are sufficient to keep the electrons within the ion beam to compensate its space charge, see Fig. 2.

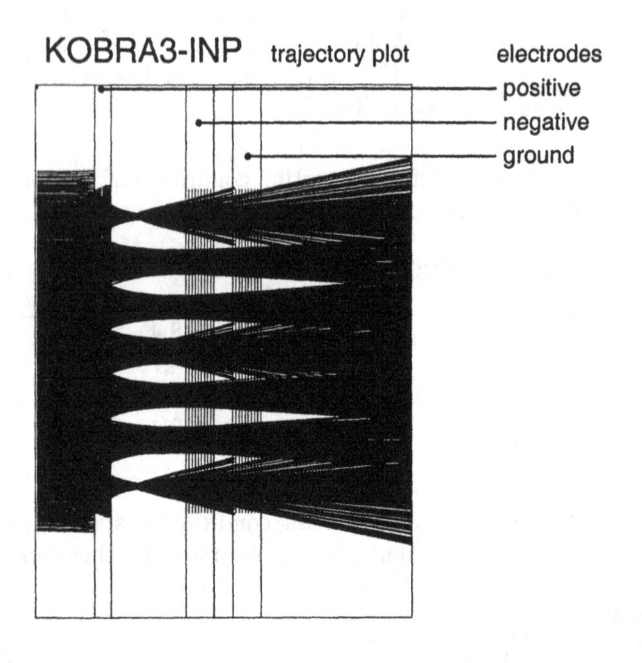

Figure 3: Extraction system with projection of all trajectories into the plane of drawing. The outer beam lets do have less current density and are mismatched therefore.

The transport of such an ion beam (50 mA with 30 kV$\times\zeta$) without compensation would not be possible as can be seen in Fig. 5. To simulate the effect of different degrees of space charge compensation usually the net current is used. A better way for the simulation is to assume a certain beam plasma potential. The advantage is a flat potential distribution within the beam with a gradient at the beam edges. Such an assumption shows the peel-off effect at the beam edges and seems to be much more realistic than using the more simple net current model, see Fig. 5.

The problem in our case is the emittance of the U^{4+} beam which is with 400 π mm mrad too large for injection into the RFQ. The single gap is shown in Fig. 6. An emittance growth takes place during acceleration most likely because the space charge distribution is not homogeneous. The location and the shape of the plasma boundary between the space charge compensated beam and the fully decompensated beam within the acceleration section is again critical for a re-distribution of the ions within phase space.

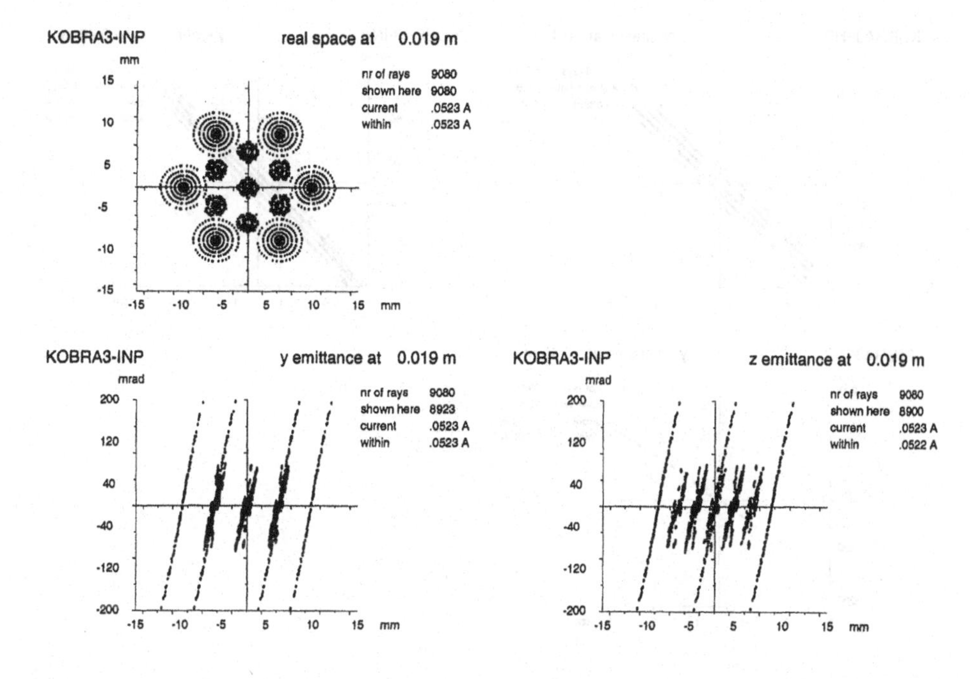

Figure 4: Simulated emittance of the extracted beam. The ion temperature has been neglected so far because its contribution to the emittance is negligible.

Table 4: Simulated emittance along the beam line in [mm mrad]

		extracted	accelerated	end of beam line (U^{4+})
rms:	y	770	350	360
	z	760	380	360
100%:	y	3200	1100	750
	z	3100	970	920
mA		55	24.2	16.4

According to Liouville's theorem the emittance should shrink by a factor of 2.08 for the full beam. This factor can be found for the rms emittance, but the beam current is reduced by a factor of 2 at the same time. All numbers for the emittances at various places are shown in Tab.4. The abberation of the beam shown in Fig. 8 indicates the presence of non linear forces, which should be able to compensate with a better shaping of the electrodes.

The bad curvature of the beam plasma boundary, see Fig. 7, is at least a partial reason of the detected emittance growth.

Like in the extraction system the negative screening electrode preserves the space charge compensation by shielding the positive potential of the acceleration

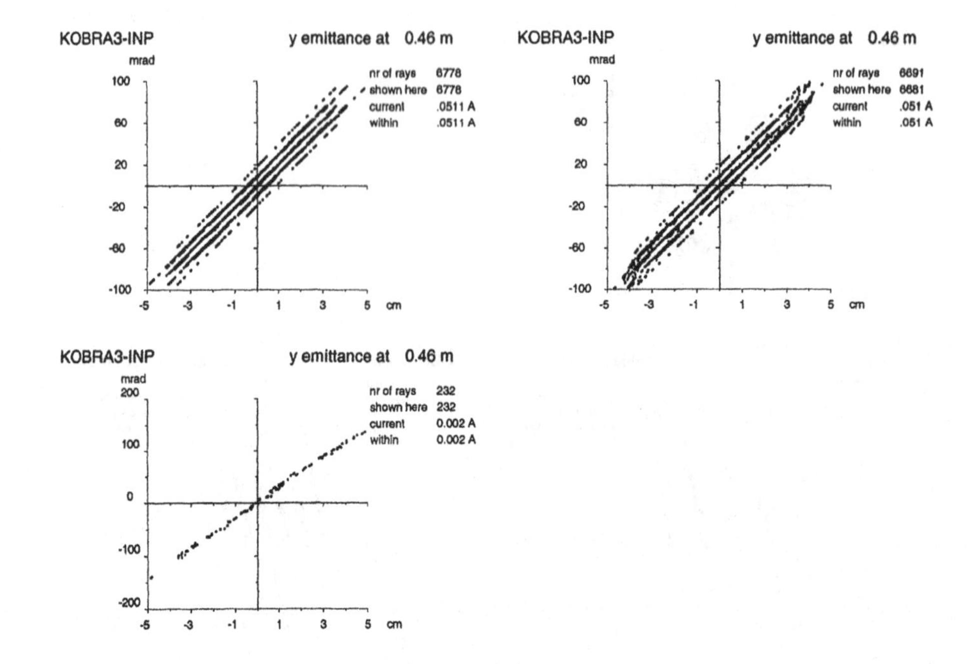

Figure 5: Emittance of the extracted beam for different degree of space charge compensation. Top left: 100% space charge compensation (scc) with 51.1 mA; top right: 99.3% scc with 51 mA; bottom left: no scc with 2 mA transported current only. Note that the hole structure has disappeared, but only 4 % of the beam could be transported.

gap against the electrons trapped within the beam. The necessary negative potential on axis depends on the electron temperature of the electrons trapped by the space charge potential of the beam. A similar electrode is necessary at the RFQ entrance.

The magnetic triplet is the first element in the beam line which has a selective influence on the ion beam. Different charge states are treated individually as can be seen in Fig. 8.

4. Conclusion

The available models to describe the physic of the ion beam seems to be correct. Using these models it is possible to describe the experimental results. From the simulations it can be concluded that the beam has to be space charge compensated to a very high degree in all field free sections of the beam transport sections. Full space charge force is acting within the extraction and the acceleration system, only. At all boundaries between compensated particle drift and uncompensated particle acceleration a nonhomogeneous distribution if ions should be avoided,

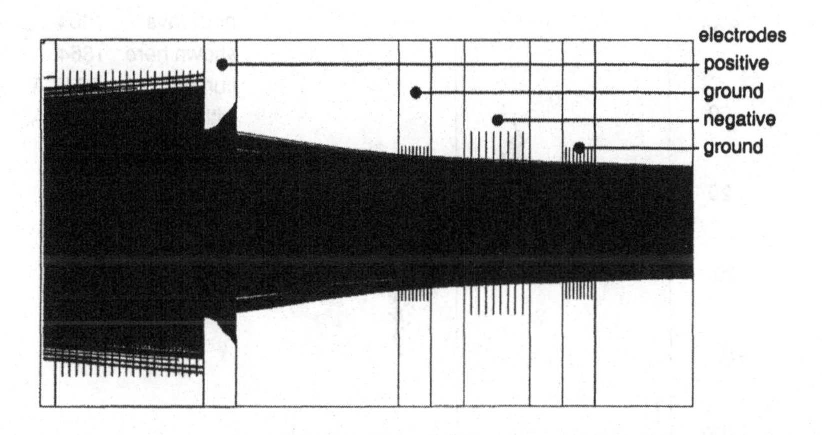

Figure 6: Single gap acceleration with screening electrode capsuled by two ground electrodes.

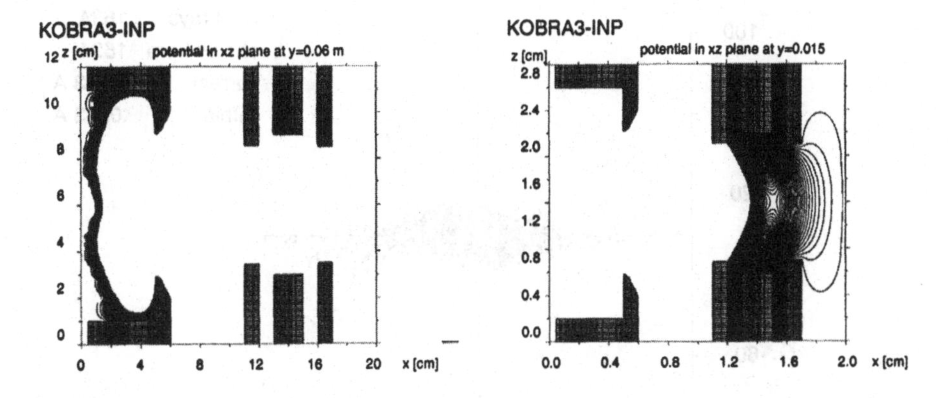

Figure 7: Left: plasma boundary. A non homogeneous distribution has been used in this example to demonstrate the reason for an emittance growth; right: negative screening potential.

otherwise emittance growth will take place because of the non linear forces. Some information, like for example the energy distribution of all particles within the plasma is still missing. But even at that state it should be possible to optimize the weak points in our beam line to reach the design value for the current in the given acceptance, which is a better electrode shaping of the acceleration electrodes

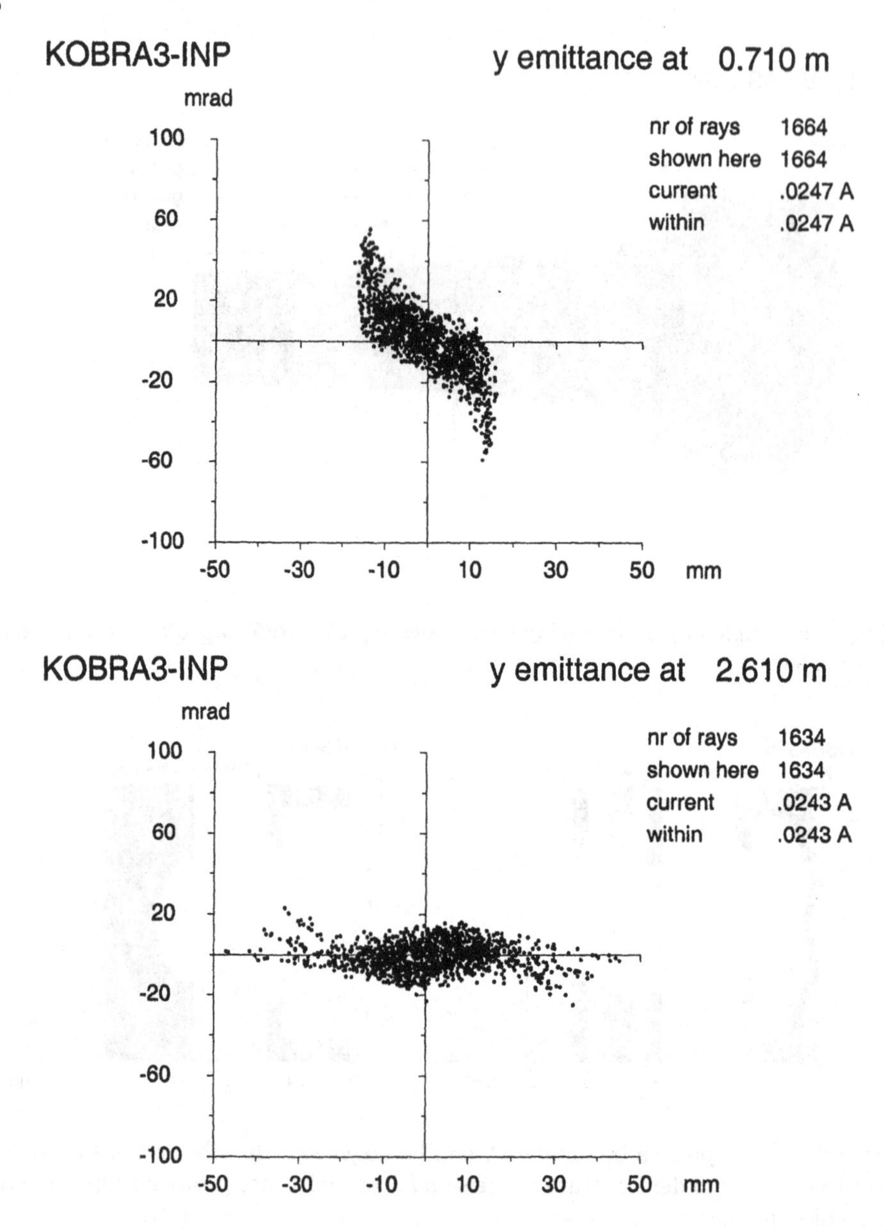

Figure 8: Top: emittance of the accelerated beam; bottom: emittance of the beam behind the triplet.

and the replacement of the triplet by a triplet with larger aperture or by a solenoid.

References:

1. Beam Intensity Upgrade of the GSI Accelerator Facility, A Technical Status Report; GSI-Report 95-05, 1995.
2. W. Barth, L. Dahl, J. Klabunde, C. Mühle, P. Spädtke; High-Intensity Low Energy Beam Transport Design Studies for thr New Injector Linac of the UNILAC.
3. R. Keller; Multi Charged Ion Production with MUCIS, GSI scientific report, 1987, p 360.
4. P. Spädtke; U^{4+} Mevva source & outlook for HIF sources; Nuclear Instruments and Methods in Physics Research A 464 (2001), p 388-394.
5. H. Reich, P. Spädtke, E. Oks; Metal Vapor Vacuum Arc Ion Source Development at GSI; Review of Scientific Instruments, Vol. 71, No. 2, Feb 2000.
6. INP, Junkernstr. 99, 65205 Wiesbaden, Germany
7. R. Hollinger, M. Galonska, F. Heymach, P. Spädtke; Uran Strahlzeit im Dezember 2001 mit der Mevvaquelle; GSI Report 2002-2, in German.
8. G. Yushkov, A. Anders, E. Oks, Ian G. Brown; Ion Velocities in Vacuum Arc Plasmas; Journal of Applied Physics, Vol. 88, No. 10, Nov 2000.

PRODUCING OF GAS AND METAL ION BEAMS WITH VACUUM ARC ION SOURCES

A.S. Bugaev, V.I. Gushenets, A.G. Nikolaev, E.M. Oks, K.P. Savkin, P.M. Schanin, G.Yu. Yushkov, I.G. Brown*

High Current Electronics Institute Russian Academy of Sciences
4 Academichesky ave., Tomsk
Russia 634055

**Lawrence Berkeley National Laboratory, University of California,*
1 Cyclotron Road, Mailstop 53, Berkeley, California
USA, 94720

Abstract. The present paper deals with peculiarities of vacuum arc ion sources employed to generate beams of gas and metal ions with the controllable ratio of ions of each type in a beam. The design and parameters of the sources are presented. The mass-charge state of ion beams is studied, and the dependence of the gas ion fraction in a beam on the length of a discharge gap is analyzed. Also presented in the paper is a new version of the design of the TITAN source and its parameters.

1. Introduction

The development of ion sources based on cold-cathode arc discharges is motivated by possibilities of their wide application in technologies for modification of surface properties of different materials and as high-current injectors for heavy ion accelerators [1-3]. The sources of such a type developed at High Current Electronics Institute provide generation of intensive pulse-periodic beams of gas and metal ions. These sources are characterized by relative simplicity of their design, convenient operation, reliability, and sufficiently long lifetime.

The TITAN-type sources feature the possibility of generating both gas ion and metal ion beams, which is provided by the use of two types of arc discharge – a constricted arc and a vacuum arc – in one discharge system. This opens prospects for producing gas-metal compounds with high strength characteristics in the surface layer of the target. To generate gas ions, use is made of a constricted cold-cathode arc discharge. In a discharge of this type, ions of the cathode material are formed in the cathode region

E. Oks and I. Brown (eds.),
Emerging Applications of Vacuum-Arc-Produced Plasma, Ion and Electron Beams, 79–90.

and plasma is created in the anode cavity through ionization of the working gas. Metal ions are generated during the operation of the cathode spots of a vacuum arc with a cathode made of the required material [4,5].

The first prototype of the TITAN source was designed at the Institute of High Current Electronics, SB RAS, more than ten years ago [6,7] and found its application in research works in Russia and abroad. Over the next years, several modifications of the source, each having its own features, were developed. Experience gained with the TITAN sources and active international scientific collaboration have made it possible to modify a number of ion sources of the Mevva type [8,9], which are based on vacuum arc discharges, in different foreign laboratories. The present paper considers the principle of operation of the TITAN sources, their characteristics and fields of application, and describes the modification of the Mevva source.

2. Principle of operation of the TITAN source

The electrode system of the TITAN ion source [4,6,10] is shown in Fig.1. Cold Mg cathodes *1* of the constricted arc are mounted one facing the other on a magnetic core in the discharge chamber in which the working gas is supplied. In the magnetic core, the magnetic flux is created by permanent Sm–Co magnets. These cathodes and loop anode *2* located between them form a Penning cell. Ferromagnetic insert *4* with a constriction channel is set in intermediate electrode *3*. Cathode *5* of the vacuum arc is fixed at the outlet of the constriction channel. The ferromagnetic insert is the magnetic conductor of a magnetic field of "arch" configuration, which is induced by circular permanent magnet *6*. Hollow cathode *7* common to both discharges is covered on its emission face with a fine emission grid. Ions are extracted with a dc accelerating voltage of 10 – 80 kV which is applied between the hollow anode kept at a high positive potential and grounded acceleration electrode *9*. To cut off the secondary electrons knocked out by the ion beam from the collector, use is made of cutoff electrode *10* to which a negative voltage of several kilovolts is applied. The electrodes *9* and *10* consist of tungsten grids of transparency ~95%. One of the versions of the source uses transverse extraction of a ribbon ion beam through a slit in the lateral face of the hollow anode [11].

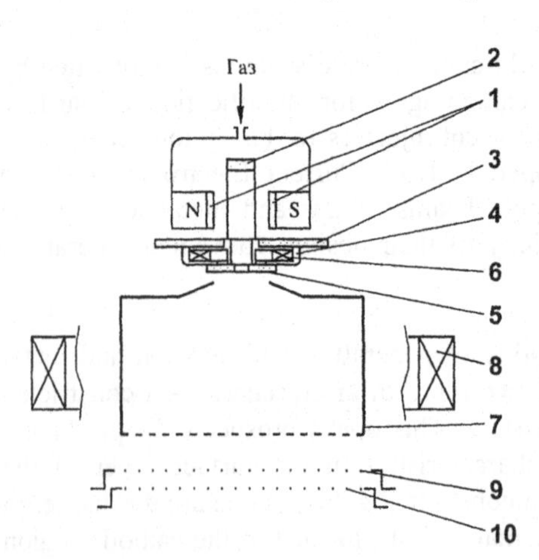

Figure 1. Electrode system of the TITAN ion source: *1* – cathodes of the constricted arc; *2* – loop anode; *3* – intermediate electrode; *4* – ferromagnetic insert; *5* – cathode of the vacuum arc; *6* – permanent Sm–Co magnets; *7* – hollow anode; *8* – solenoid; *9* – acceleration electrode; *10* – cutoff electrode

When operated in the pulsed mode, the TITAN ion source requires stable operation of the vacuum arc for a long time, since this process is responsible for the time the device operates continuously. Initiation of the vacuum arc by a low-current gas discharge in the TITAN source has made it possible to increase the continuous operation time of the ion source to a value determined only by the time of cathode material erosion occurring during the operation of the vacuum arc. A modified version of the TITAN source, which operates in the steady-state conditions, has also been developed [12]. This version provides generation of wide-aperture beams of gas and metal ions with currents of 100–300 mA.

All versions of the TITAN source use magnetic fields of different configurations. Permanent magnets located on the cathodes of the Penning cell or along the perimeter of the discharge chamber, as in the TITAN-2 source [13-15], provide for initiating an auxiliary Penning discharge. In addition, the magnetic field improves the parameters and emissivity of the source. So, the magnetic field of "arch" configuration in the cathode region of the vacuum arc stabilizes the motion of the cathode spot on the cathode surface [16]. This promotes more stable operation of the source and more uniform wear of the cathode. A magnetic field of up to 10 mT induced by solenoid *8* in the anode region the anode region increased substantially the efficiency of extraction. This is due to the fact that when the polarity of the potential of the near-anode layer is changed, the ion current "switches" to the emission electrode [17,18]. Moreover, the magnetic field changes the ion current density distribution in the cross section of the beam [4, 16].

A magnetic field of up to 1 T induced in the cathode region of the discharge can provide an increase in metal ion charge by a factor of 1.5–2, as in the TITAN-2 source [19].

3. Production of gas and metal ions in the TITAN source

A constricted discharge is initiated by applying a high-voltage pulse of duration ~20 μs between cathodes *5* and hollow anode *8*. Upon initiation of the auxiliary discharge and application of a pulsed voltage to these electrodes, a constricted arc of current 20–60 A and duration 400 μs is ignited between them. The operation of the discharge and the simultaneous flow of the working gas through the constriction channel result in more efficient plasma production. The gas discharge plasma fills the hollow anode and ions are extracted from the developed plasma surface, which is stabilized by the emission grid.

If voltage is applied between the hollow anode *8* and cathode *6* upon initiation of the auxiliary discharge, a vacuum arc is initiated between these electrodes and cathode spots are formed at the cathode surface. The arc current is 40–150 A and the pulse duration is 400 μs. During the operation of the arc, the plasma flux of the cathode material coming from the cathode spots fills the hollow anode and a metal ion beam is formed in the same manner as a gas ion beam is generated from the plasma of a constricted discharge. The pulse repetition rate of the discharges and, consequently, that of the ion beam is controlled in the range from 10 to 50 Hz. Because the discharges have independent power supplies and the processes of plasma formation in the

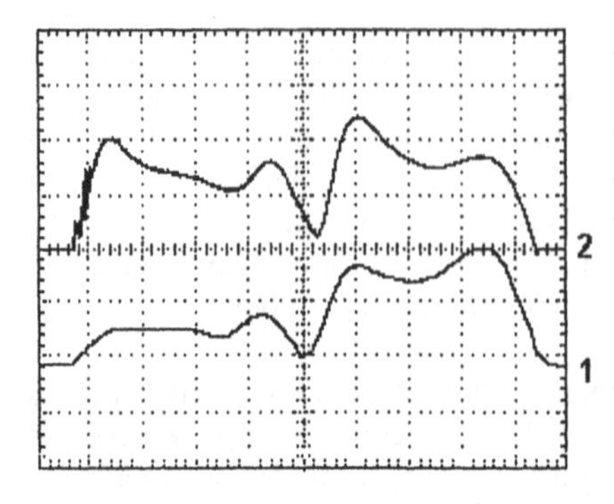

Figure 2. Oscillograms of the pulses produced by the TITAN-3 source: *1* - discharge current I_d (on the left – constricted arc, on the right – vacuum arc); *2* – ion current I_i (on the left – N_2 ions, on the right – vacuum arc). Scale: τ= 100 μs/div., I_d= 40 A/div., I_i= 0.2 A/div. U_{accel}= 50 kV.

discharges are different and independent of each other, the simultaneous operation of both discharges can provide a two-component gas-metal ion beam (Fig. 2). In this case, the fraction of ions of each component is easily varied by varying the discharge current [20]. The current of the ion beam extracted from the plasma ranges from 0.1 to 0.3 A for gas ions and from 0.2 to 0.5 A for metal ions.

To determine the content of metal ions in the case where only a gas discharge is operative and the content of gas ions in the case where a metal ion beam is generated, the mass-charge spectrum of the ion beam was measured when the ion source was operated in these two limiting modes [20]. At the early stage of the vacuum arc initiation, the constitution of the ion beam could vary during the characteristic time of 100 μs. The average charge of the metal ions in the beam was higher and, at the same time, an impurity of H^+ ions was present whose percentage decreased from 10% to zero within the mentioned time. Later on, the constitution of the ion beam remained practically unchanged, and this may be evidence of the stabilization of the processes that are responsible for the plasma production in the vacuum arc.

A typical pressure dependence of the beam constitution for the emission of ions from the plasma of a vacuum arc operating on a titanium cathode is given in Fig. 3. It can be seen that as the pressure is increased from $2 \cdot 10^{-5}$ to $8 \cdot 10^{-4}$ Torr by increasing the rate of supply of nitrogen to the discharge chamber, the fraction of Ti+ ions increases, while the fractions of higher charged ions, Ti^{2+} and Ti^{3+}, decrease. As the pressure reaches $2 \cdot 10^{-4}$ Torr, the content of Ti^{3+} ions decreases practically to zero. It should be noted that the average charge of titanium ions in the test pressure range is 1.4–1.8, which is lower than the values characteristic of ion sources operating at lower residual gas pressures.

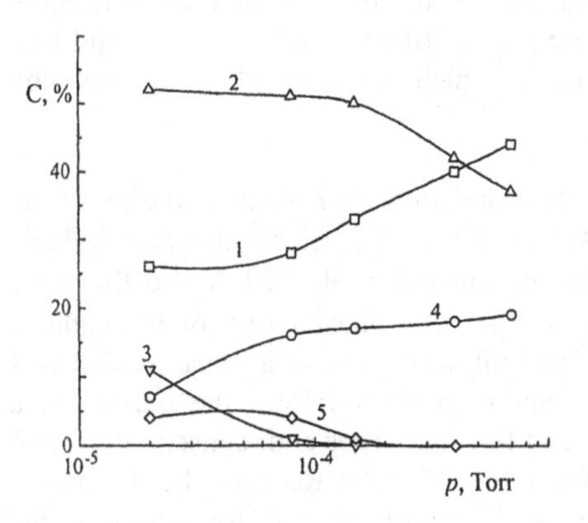

Figure 3. Constitution of a titanium ion beam *versus* nitrogen pressure: *1* – Ti^+; *2* –Ti^{2+}; *3* – Ti^{3+}; *4* – N_2^+; *5* – N^+; I_b = 60 A; U_{accel} = 21 kV; B = 3 mT.

From the above dependence it follows that the degree of "pollution" of the beam with nitrogen ions (N_2^+ and N^+) is 10–20%, and as the pressure is increased, the fraction of N_2^+ molecular ions increases, while that of N+ atomic ions decreases up to their disappearance at a pressure of (3–6)·10^{-4} Torr. A similar effect of pressure on the beam constitution was observed for other metal ion types and for other gases, except, naturally, inert gases. For Ar and Kr fill gases, double-atom ions were absent.

An axial magnetic field of up to 10 mT, which was applied to the anode region of the discharge for increasing the ion emission current from the vacuum arc plasma, did not change the average charge of the metal ions. At the same time, as the magnetic field was reduced to zero the degree of pollution of the beam with gas ions decreased from 10–20 to 5–10%. Therefore, the subsequent versions of TITAN ion source did not use an axial magnetic field.

When the plasma of a constricted discharge generated ions, there were no ions of the materials of the vacuum arc and constricted discharge. In the case where the working gas was nitrogen or oxygen, the beam contained molecular and atomic ions of these

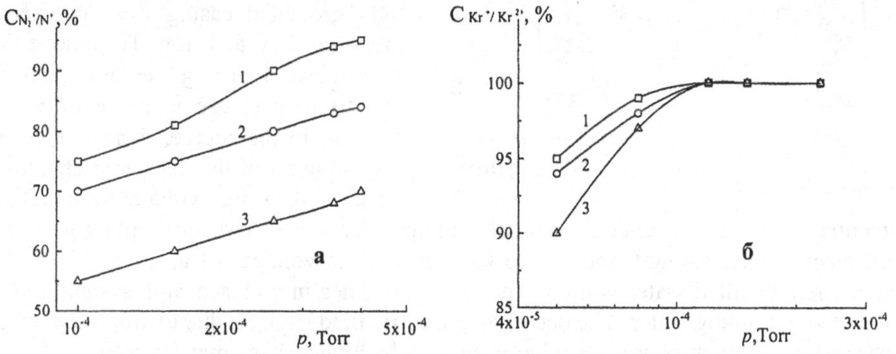

Figure 4. N^+ to N_2^+ (*a*) and Kr^{2+} to Kr^+ (*b*) ion content ratio *versus* pressure: *1* – B = 0 mT; *2* – 3 mT; *3* – 6 mT; I_c = 20 A; V_{accel} = 21 kV (for both figures)

gases. Their content ratio depended on the pressure and on the magnetic field applied to the discharge gap. In the dependence presented in Fig. 4a it can be seen that an increase in pressure or in magnetic field increases the molecular ion fraction. In a krypton ion beam, a similar dependence for which is shown in Fig. 4b, only singly and doubly charged ions were observed. As the pressure was increased from 5·10^{-5} to 10^{-4} Torr, Kr^{2+} ions disappeared, and the effect of the pressure and magnetic field on the Kr^{2+} ion fraction was similar to the effect of these factors on the atomic ion fraction in the case where nitrogen and oxygen ion fractions were generated.

4. The TITAN-3 ion source

The design of the last modification of the TITAN source is shown schematically in Fig. 5. Despite the fact that the principle of operation, the parameters, and the system of ion beam extraction of the source are the same as those of the base model [6, 16], a number of units and parts of the source have been changed substantially [21, 22]. So, the TITAN-3 source employs gas insulation to preclude the presence of carbon and

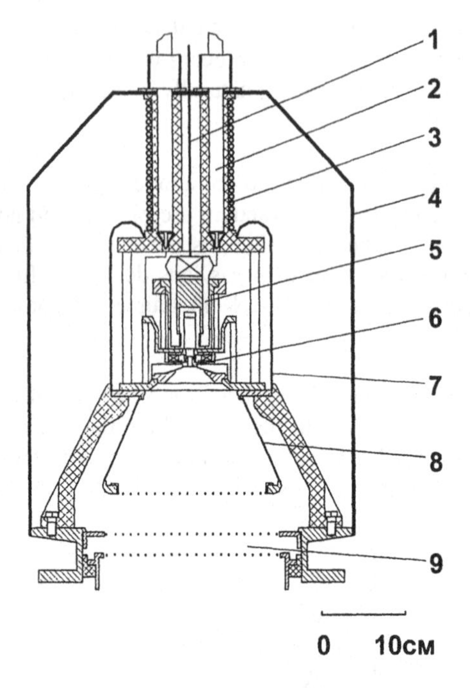

Figure 5. Design of the TITAN-3 ion source. *1* – pipe for supplying the working gas to the discharge chamber; *2* – high-voltage cable; *3* – pipe for supplying cooling water to the discharge chamber; *4* – grounded casing; *5* – cathodes of the constricted arc; *6* – cathode of the vacuum arc; *7* – high-voltage screen of the discharge chamber; *8* – hollow anode; *9* – acceleration and cutoff electrodes.

carbide compounds in the ion-doped layer of the treated specimens, which takes place when a system of oil insulation and cooling of the discharge chamber by oil diffusion through vacuum seals is used [23]. To this end, dry air or nitrogen is supplied to air-tight grounded casing *4* at an excess pressure of up to 1 atm. To remove the heat released during the operation of the discharges, use is made of water cooling. In the source, there is electric decoupling *3* of the discharge chamber, which is at a high-voltage accelerating potential, and of the grounded supply of cooling water. The electric decoupling provides sufficient electric strength and low leakage currents at voltages of up to 80 kV in the case where distilled water is used. The latter circulates in a closed pipe system and is cooled with running water. The decrease in electric filed strength due to amplification at sharp edges of the discharge chamber is ensured by high-voltage metal screen *7*.

In all, more than ten TITAN ion sources of various modifications have been designed. They are now being operated at the Institute of High Current Electronics, Dalian and Guangzhou Universities (China), and at the Andrzej Soltan Institute for Nuclear Studies (Poland) where they are mainly used to modify the surface properties of materials by ion implantation. With the use of these ion sources gas-metal compounds of required stoichiometry are produced in the near-surface layer of the treated articles. Thus, as nitrogen and titanium ions were implanted together into an α-iron target, TiN particles of volume percentage of up to 10% were formed in the surface layer of the target. This resulted in a substantial dispersion hardening of the alloyed layer [24]. For the implantation of nitrogen and titanium into titanium or molybdenum articles, TiN and Mo_2N particles were formed in the near-surface layer, and as the irradiation dose was increased, solid layers of the embedded phase showing a very high adhesive strength were formed [25, 26]. With high-dose implantation of carbon, nitrogen, and silicon into molybdenum, the formation of molybdenum carbides, nitrides, and silicides was attained, resulting in the appearance of an amorphous molybdenum phase [24, 26]. When Cu, Mo, and Pb ions were implanted into molybdenum, a developed high-density dislocation structure formed in the near-surface layer and, in addition, the initial stage of molybdenum creep shortened [25]. These and some other studies in physics of metals performed with the use of TITAN ion sources are described in detail elsewhere [23, 27, 28].

The TITAN ion source has also been applied to improve the operating parameters of the electrodes of an electric-discharge laser, to harden cutters, stamps, drills, and cutting tools for increasing their lifetime and to improve the serviceability of other articles and tools.

An interesting application of the ion source has been found at Zhengzhou University (China) where it is used as a tool for perforation of cells of different biological objects in genetic engineering.

5. Modification of sources of the Mevva type

Experience gained with the TITAN sources has allowed design improvements of vacuum-arc-based ion sources of the Mevva type. The main idea of this improvement was to induce an axial magnetic field in the discharge gap of the source. The initiation of a vacuum arc by a Penning discharge in a strong magnetic field was employed to modify the Mevva-V vacuum-arc ion source [9, 29]. To this end, solenoid *4*, auxiliary Penning anode *2*, and a system of supplying the working gas was added to the design of the source (Fig. 6). The above peculiarity made possible an increase in the lifetime of the Mevva ion source to 10^7 pulses.

The use of only a strong magnetic field without supplying the working gas in sources of the Mevva type results in an increase in ion beam average charge and in the appearance of high-charge ions [8, 30]. So, for instance, with no magnetic field, the average charge of ions of the vacuum arc ranges from 1 (for carbon) to 3.08 (for tungsten). When a magnetic field is applied, the average charge of the ion beam increases by a factor of 1.2–2.5, depending on the cathode material [31, 32]. The dependence of the charge distribution on the value of the magnetic field tends to a limit at 1 T [30].

Another result of the modification was the possibility of producing gas ions along with metal ions in the Mevva source [33, 34]. Despite the fact that the change in the type of working gas and cathode material determined some peculiar effects of this gas, in all cases the increased pressure provided a decrease in the current of ions extracted from the plasma, a decrease in the multiply

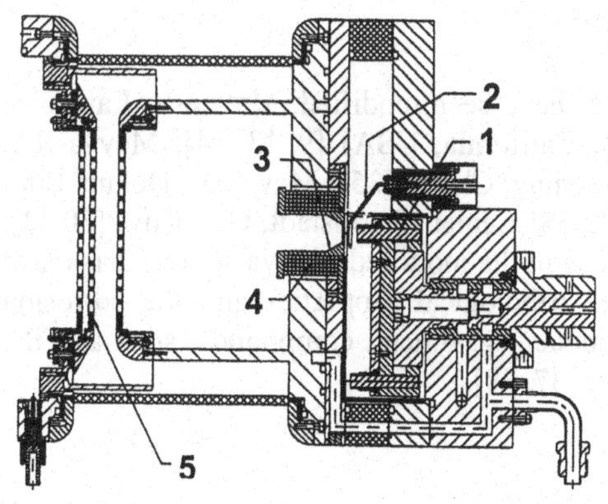

Figure 6. Modified Mevva-V source with a Penning system of initiation: *1* – cathode; *2* – auxiliary anode; *3* – anode; *4* – solenoid; *5* – multi-aperture extraction system

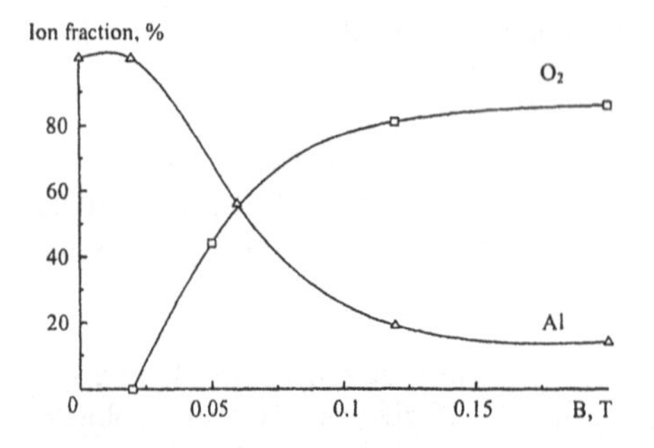

Figure 7. Fraction of Al ions (total Al^+, Al^{2+} and Al^{3+}) and O_2 ions (total O_2^+ and O^+) in the beam as a function of magnetic field. p= $2 \cdot 10^{-4}$ torr.

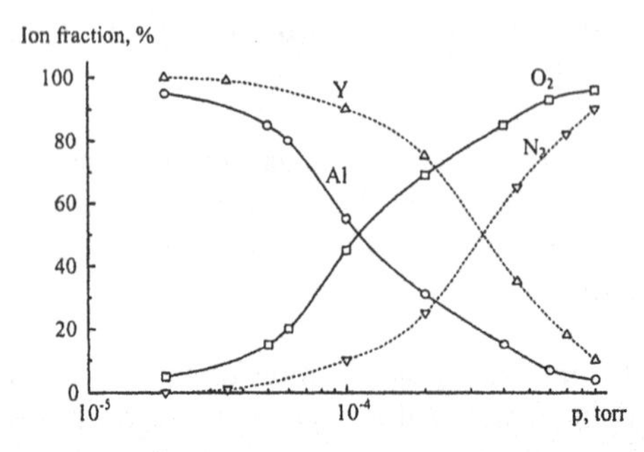

Figure 8. Fraction of Al / O_2 ions (solid lines) and Y / N_2 ions (dash lines) in the beam as a function of system pressure. B= 0.1 T.

charged ion fraction, and the appearance of gas ions in the mass-charge constitution. The effect of the gas pressure on the mass-charge ion distribution manifest itself even at minimum values of pressure and multiply charged ions therewith turn out to be more susceptible to this effect.

However, experiments have shown the effective generation of the leaked-in gas ions and the emergence of their appreciable fraction in the ion spectrum to be possible only upon achieving some critical value by the magnetic field. Thus, the magnetic field dependence of the mass-charge state of the ion beam shows that a magnetic field of value higher than 20 mT is required to generate O ions in the vacuum arc plasma in the case of oxygen leak in a system with an Al cathode (Fig. 7). The change in the value of the magnetic field in the range from 20 mT to 0.2 mT makes it possible to control the ratio of the atomic percentage of O ions in the beam from nearly 0 % to 85 % and that of Al ions from 100 % to 15 %.

Varying the value of the gas pressure at a constant magnetic field can also do the control of the ratio of the gas ion fractions and those of metal ions in the beam. It can be seen from the dependencies shown in Fig. 8 that the ratio of the O ion percentage for the case of the vacuum arc with the Al cathode ranges from 4 % to 95 %, that of the Ni ion percentage for the case of an Y cathode is from 0 % to 90 % as the pressure is increased from $2 \cdot 10^{-5}$ Torr to 10^{-3} Torr.

For the last five years, the following sources have been modified: Mevva-V, Lawrence Livermore National Laboratory (Berkeley, California, USA) [9, 13, 14]; Mevva-IIA, Institute of Low-Energy Nuclear Physics (Beijing, China) [35]; Mevva-V. Dokuts Eilul University (Izmir, Turkey) [36]; Mevva-IV, GSI Center (Darmstadt, Germany) [30,33]. The possibility to generate gas and metal ions in the modified Mevva sources has made it possible to use them for improving the tribological properties and the corrosion characteristics of articles and tools by producing gas-metal compounds, such as TiN, PtN, AlO, and ZrO, in the near-surface layer [37, 38].

6. Processes leading to the formation of gas ions in a vacuum arc discharge

To find out where in the discharge gap gas ions are generated, experiments have been performed on the effect of the discharge gap length on the gas and mass-charge states of the ion beam. To this end, additional moving anode *3* (Fig. 9) was set in the source of the Mevva type, and the voltage was applied to main hollow anode *5* through the resistance R1 of value several kΩ. The mass-charge state of the ion beam was analyzed by a time-of-flight mass spectrometer.

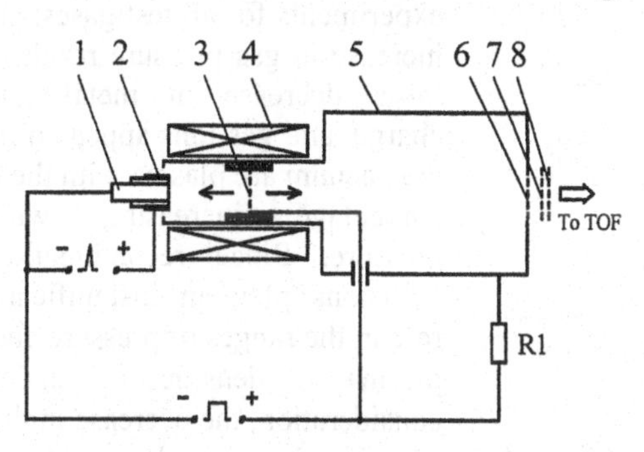

Figure 9. Experimental vacuum arc source with moving anode and means for gas injection. 1- cathode, 2- trigger, 3- moving anode, 4- coil, 5- hollow anode, 6- emission grid, 7,8- accel-decel system.

The experiments have shown that at a pressure of the order of 10^{-4} Torr the increase in the "cathode- anode" spacing causes a substantial increase in the gas fraction present in the metal ion beam even in the absence of the magnetic field [21]. The dependencies of the gas ion fraction in the metal ion beam on pressure are presented in Fig. 10 for a varying distance between the cathode and the anode. It can be seen from this figure that by varying the discharge gap length from 0.2 cm to 7 cm it is possible to increase the ratio of the atomic percentage of O ions in the Al ion beam from less than 20 % to 70 %. In the case of a Ti cathode the percentage of N ions also increases from 12 to 92 %.

When changing the cathode – anode distance, the gas ion fraction first increases linearly and reach the highest value at a distance of 3 – 8 cm (Fig. 11). Further there occurs "saturation" of this dependence and the gas ion fraction becomes constant. Thus, it is possible to conclude that gas ions are generated near the vacuum arc cathode at a distance of the order of several cm and, further, cooling of the vacuum arc electrons

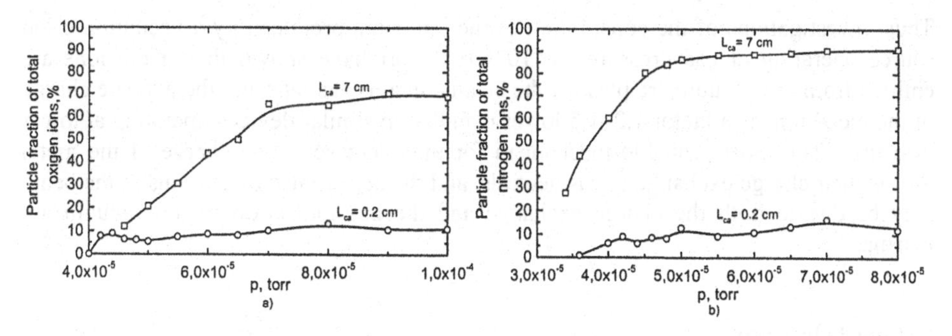

Figure 10. Particle fraction of gas ions in the total ion beam as a function of system pressure. B= 0. a) Cathode – Al, gas – O_2; b) cathode - Ti, gas – N_2.

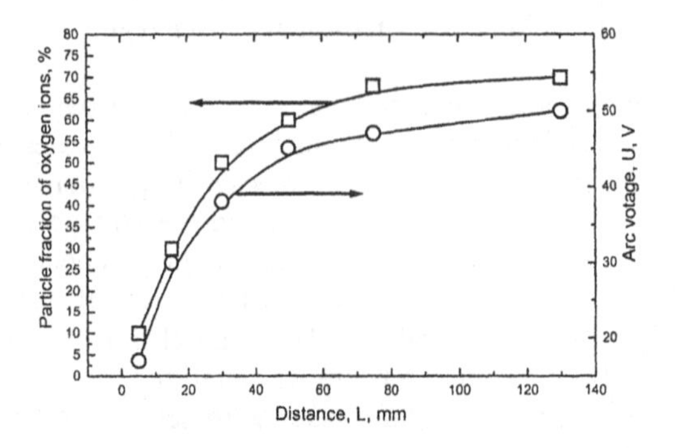

Figure 11. Particle fraction of ion beam and arc voltage as a function of cathode-anode distance. Cathode – Al, gas – O_2, B= 0, p = 1.2 10^{-4} Torr.

takes place and gas ionization terminates.

As mentioned above, the same tendency was observed in all experiments for all test gases: an increase in gas pressure resulted in a decrease in metal ion charge, and gas ions appeared in the vacuum arc plasma with their percentage increasing with pressure. Since recombination collisions play an insignificant role in the ranges of pressure and plasma densities under consideration, the decrease in the charge of metal ions is, most probably, due to their stepwise charge exchange at gas ions (molecules). A detailed investigation of the effect of pressure on the constitution and ion charge states of the plasma of a vacuum arc is described in [39].

It is well known that a magnetic field of up to 10 mT applied to the plasma of a vacuum arc magnetizes the plasma electrons, and the ions coming from the vacuum arc cathode spots are focused by the self-consistent electric field of the plasma [17]. Provided that the formation of gas ions during the operation of a vacuum arc is due to charge exchange processes, it can be suggested that gas ions, in contrast to metal ions, have no directed velocities and move chaotically. The fraction of the ions extracted as a beam will then be proportional to the ratio of the area of the plasma emission surface to that of the plasma surface bordering on the other electrodes of the discharge system. The application of a magnetic field has the result that the gas ions are confined by the negative potential of the plasma, thus being prevented from going away to the electrodes of the discharge system, and so the percentage of gas ions in the extracted beam increases.

Thus, investigations of the constitution of the ion beam produced by a vacuum-arc ion source operating at gas pressures of 10^{-4}–10^{-3} Torr have shown that if the ions are emitted from the vacuum arc plasma, the beam contains gas ions and the average charge of the metal ions is a factor 1.2–1.5 lower than that in similar devices operating at lower pressures. The most probable mechanism for the decrease in the charge of the metal ions is their charge exchange at gas neutrals, and the appearance of gas ions in the beam may be due to both the charge exchange and the gas ionization in the vacuum arc plasma.

Acknowledgments

The work on design improvement and on some applications of the TITAN sources was supported by contracts with the Lawrence Berkeley National Laboratory, California,

USA, under the program IPP Trust-1 and by the Ministry of Science program No 0.06.01.0103.

References:

1. I.G. Brown (Ed.) *The physics and technology of ion sources*. - Wiley, New York, 1989.
2. B.H. Wolf (Ed.) *Handbook of ion sources*. – Boca Raton, Fl.: CRS Press, 1995.
3. E.M. Oks, *Rev. Sci. Instrum.* **69,** 776 (1998).
4. S.P. Bugaev, A.G. Nikolaev, E.M. Oks, P.M. Schanin, and G.Yu. Yushkov, *Rev. Sci. Instrum.* **65**, 3119 (1994).
5. .G. Brown, *Rev. Sci. Instrum.* **63,** 2351 (1992).
6. S.P. Bugaev, A.G. Nikolaev, E.M. Oks, P.M. Schanin, and G.Yu. Yushkov, *Rev. Sci. Instrum.* **63**, 2422 (1992).
7. E.M. Oks, and G.Yu. Yushkov, *Proc. Mevva workshop'95*, Berkeley, USA, 1995, p. 24.
8. A. Bugaev, V. Gushenets, A. Nikolaev, E. Oks, G.Yu.Yushkov, A. Anders, and I.G. Brown, *Proc. 18th Intern. Symp. on Discharges and Electrical Insulation in Vacuum.* Eindhoven, Netherlands, 1998, V. 1, p. 256.
9. A.G. Nikolaev, G.Yu. Yushkov, E.M. Oks, R.A. MacGill, M.R. Dickinson, and I.G. Brown, *Rev. Sci. Instrum.* **67**, 3095 (1996).
10. A.G. Nikolaev, E.M. Oks, P.M. Schanin, and G.Yu. Yushkov, *Proc. Vth Intern. Conf. on Ion Sources.* Beijing, China, 1993, p. 56.
11. A.G. Nikolaev, P.M. Schanin, and G. Yu. Yushkov, *Instruments and Experimental Techniques.* **37**, 329 (1994).
12. A.V Vizir', A.G. Nikolaev, E.M. Oks, P.M. Schanin, and G. Yu. Yushkov, *Instruments and Experimental Techniques.* **36,** 434 (1993).
13. A.G. Nikolaev, E.M. Oks, P.M. Schanin, and G.Yu. Yushkov, *Rev. Sci. Instrum.* **67**, 1213 (1996).
14. A. Nikolaev, E. Oks, G.Yu. Yushkov, R. MacGill, M. Discinson, and I.G. Brown, *Proc. 18th Intern. Symp. on Discharges and Electrical Insulation in Vacuum.* Eindhoven, Netherlands, 1998, V. 1, p. 101.
15. A.G. Nikolaev, G.Yu. Yushkov, P.M. Schanin, and E.M. Oks, *Proc. 22th Intern. Confer. on Phenomena in Ionized Gases.* Hoboken, USA, 1995, V. 1, p. 95.
16. S.P. Bugaev, A.G. Nikolaev, E.M. Oks, P.M. Schanin, and G.Yu. Yushkov, *Proc. of 15th Intern. Symp. on Discharges and Electrical Insulation in Vacuum.* Darmstadt, Germany, 1992, p. 686.
17. A.G. Nikolaev, E.M. Oks, P.M. Schanin, and G.Yu. Yushkov, *Zh. Tekh. Fiz.* **62,** 140 (1992).
18. A.I. Ryabchikov, *Izvestiya Vuzov. Phys.* **3,** 34 (1994).
19. A. S. Bugaev, V. I. Gushenets, G. Y. Yushkov, E. M. Oks, A. Anders, I. Brown, A. Gershkovich, and P. Spadke, *Russian Physics Journal.* **44,** 912 (2001).
20. A.S. Bugaev, V.I. Gushenets, A.G. Nikolaev, E.M. Oks, and G.Yu. Yushkov, *Izvestiya Vuzov. Phys.* **2,** 21 (2000).
21. A.S. Bugaev, V.I. Gushenets, A.G. Nikolaev, E.M. Oks, and G.Yu. Yushkov, *Proc. Ist Intern. Congr. on Radiation Physics, High Current Electronics and Modification of Materials.* Tomsk, Russia, 2000, V. 3, p. 204.
22. S. P. Bugaev; A. G. Nikolaev; E. M. Oks; G. Y. Yushkov; P. M. Schanin; and I. Brown, *Russian Physics Journal.* **44,** 921 (2001).
23. A.N. Tyumentsev, Yu.P., Pinzhin, A.D. Korotaev, A.F. Safarov, S.P. Bugaev, A.G. Nikolaev, and G.Yu. Yushkov, *The Physics of Metals and Metallography.* **83,** 179 (1997).
24. A.D. Korotaev, A.N. Tyumentsev, Yu.P., Pinzhin, O.V. Panin, A.F. Safarov, S.P. Bugaev, P.M. Schanin, and G.Yu. Yushkov, *Surface and coating technology.* **96,** 89 (1997).
25. A.N. Tyumentsev, Yu.P., Pinzhin, A.D. Korotaev, A.D. Bekhert, A.O. Savchenko, Yu.R. Kolobov, P.M. Schanin, and G.Yu. Yushkov, *The Physics of Metals and Metallography.* **9,** 123 (1992).
26. A. N. Tyumentsev, Yu.P. Pinzhin, A.D. Korotaev, A.E. Behert, A.O. Savchenko, Yu.R. Kolobov, S.P.Bugaev, P.M. Schanin, and G.Yu.Yushkov, *Nucl. Instrum. And Meth. Phys. Res.* **B 80/81,** 491 (1993).
27. A.D. Korotaev, A.N. Tyumentsev, Yu.P. Pinzhin, A.F. Safarov, S.P. Bugaev, P.M. Schanin. N.N. Koval, G.Yu. Yushkov, and A.V. Vizir, *Proc. Fall Meeting of the Materials Research Society.* Boston, USA, 1996, p. 119.

28. P.M. Schanin, N.N. Koval, D.P. Borisov, G.Yu. Yushkov, A.V. Vizir, A.G. Nikolaev, A.D. Korotaev, A.N. Tumentsev, and Yu.P. Pinzhin, In: *Materials and Process Engineering Project for SNL/NIS Industrial Partnering Program.* AM-7676, V.2, 1994.
29. A.G. Nikolaev, G.Yu. Yushkov, E.M. Oks, R.A. MacGill, M.R. Dickinson, and I.G. Brown, *Proc. 17th Intern. Symp. on Discharges and Electrical Insulation in Vacuum.* Berkeley, USA, 1996, p. 562.
30. E.M. Oks, I.G. Brown, M.R. Dickinson, R.A. MacGill, H. Emig, P. Spadtke, and B.H. Wolf, *Appl. Phys. Lett.* **67,** 200 (1995).
31. A. Anders, G.Yu. Yushkov, E.M. Oks, A.G. Nikolaev, and I.G. Brown, *Rev. Sci. Instrum.* **69,** 1332 (1998).
32. A.G. Nikolaev, E.M. Oks, and G.Yu. Yushkov, *Technical Physics.* **43,** 514 (1998).
33. P. Spadtke, H. Emig, B. Wolf, and E. Oks, *Rev. Sci. Instrum.* **65,** 3113 (1994).
34. A.G. Nikolaev, E.M. Oks, and G.Yu. Yushkov, *Technical Physics.* **43,** 1031 (1998).
35. A. Nikolaev, E. Oks, X. Zhang, and C. Cheng, *Rev. Sci. Instrum.* **69,** 807 (1998).
36. A. Oztarhan, I.G. Brown, P. Evans, G. Watt, C. Bakkaloglu, A.S. Eltas, E.M. Oks, A.G. Nikolaev, S. Selvi, Z. Tek, and I. Saklakoglu, *Proc. 11th Intern. Conf. on Surface Modification on Metals by Ion Beams.* Beijing, China, 1999, p. 66.
37. E.M. Oks, G.Yu. Yushkov, P.J. Evans, A. Oztarhan, I.G. Brown, M.R. Dickinson, F. Liu, R.A. MacGill, O.R. Monteiro, and Z. Wang, *Nucl. Instrum. And Meth. Phys. Res.* **B 127/128,** 782 (1997).
38. B.H. Wolf, H. Emig, D.M. Ruck, P. Spadtke, and E. Oks, *Nucl. Instrum. And Meth. Phys. Res.* **B 106,** 651 (1995).
39. E. Oks, and G. Yushkov, Proc. of 17th Intern. Symp. on Discharges and Electrical Insulation in Vacuum. Berkelry, USA, 1996, V. 2, p. 584.

HIGH CURRENT ELECTRON SOURCES AND ACCELERATORS WITH PLASMA EMITTERS

V. I. Gushenets, and P.M. Schanin

High Current Electronics Institute
Russian Academy of Sciences
4 Academichesky ave., Tomsk
634055, Russia

Abstract. The design and main features of vacuum arc plasma cathode electron guns for different application as well as the status of experimental research in recent years are discussed. The attractiveness of this kind e-gun is due to its capability of creating high current, broad or focused beams, both in pulsed, burst and steady state modes of operation. The mesh plasma cathode in combination with a vacuum arc used for formation of a electron beams with emission current from amperes to kiloamperes as well as beam current densities up to 100 A/cm^2, accelerating voltage from several keV to hundred keV and pulse length from 0.1 μs to 100 μs, repetition rate up to hundred kHz.

1. Introduction

The electron sources and accelerators with plasma emitter and the methods of producing pulsed electron beams and the physical processes associated with electron extraction from plasma in plasma emitters are considered. In plasma devices, the plasma source is either an arc with a cathode spot, which operates in the vapors of the expendable cathode material, or gas supplied to the discharge gap. The cathode spot exhibits almost limitless emissivity that makes possible design of electron sources and electron accelerators generating pulsed beams of currents $10 - 10^3$ A, pulse duration $10^{-7} - 10^{-3}$ s, and beam cross section up to 10^{-4} cm^2 at current densities of up to several tens of amperes per square centimeter. Plasma-emitter electron sources and accelerators based on arc discharges feature higher energy efficiency, longer lifetime, and higher emission current densities than hot-emitter accelerators and are capable of operating in a wide range of pressure ($5 \times 10^{-2} - 5 \times 10^{-4}$ Pa). Compared to explosive-emission cathodes, plasma-emission cathodes allow generation of longer current pulses with a reasonable current density distribution over the beam cross section.

E. Oks and I. Brown (eds.),
Emerging Applications of Vacuum-Arc-Produced Plasma, Ion and Electron Beams, 91–104.

2. Principle of operation of a plasma emitter

In a plasma emitter, an arc is initiated on the cathode surface by a surface discharge over an insulator, which separates the cathode and the trigger electrode, or by a current pulse, which evaporates the conductive film from the insulator surface. Such a triggering system features simplicity and sufficient lifetime and provides reliable initiation of a cathode spot. The lifetime of a triggering system is determined by the cathode material removal and by the failure of the insulator separating the cathode and the trigger electrode. The lifetime of the triggering system in question ranges to (2 – 5) $\times 10^6$ pulses at a charge of (1 – 0.6) $\times 10^{-2}$ C transferred in a pulse. In a number of the electron sources designed, use is made of up to seven triggering systems that allows a substantial increase in the lifetime of the device as a whole by decreasing the charge transferred in the circuit of one cathode. Two main designs of the triggering system employed in plasma emitters and the geometry of a plasma emitter are presented in Fig. 1. A more detailed description of a similar system for initiating a cathode spot and its various designs can be found elsewhere [1].

Upon initiation of the arc, a discharge is ignited between hollow anode *4* and cathode *1*. In experiments [2], two regimes of initiation and operation of the discharge were observed, being conditionally called the vacuum and gas regimes. A rather low operating pressure at which the free path for the ionization reaction is much longer than the electrode separation characterizes the both regimes. On exceeding some pressure P_{cr} [3], the residual gas in the discharge gap, however, exerts an appreciable effect on the discharge initiation and operation. There is a generally accepted opinion that in the vacuum regime the plasma inside the anode cavity is formed by the expansion of the cathode spot toward the anode-cathode gap. This regime is most extensively employed in metal-ion sources but it may well be applied for developing high-efficiency electron sources where stringent requirements are imposed on the vacuum conditions. A flaw of this regime of operation is the rather high instability of the electron emission current pulse. This is associated with the fact the plasma and the vapors nonuniformly get into the anode cavity due to the unstable nature of the cathode spot. Figure 2*a* shows

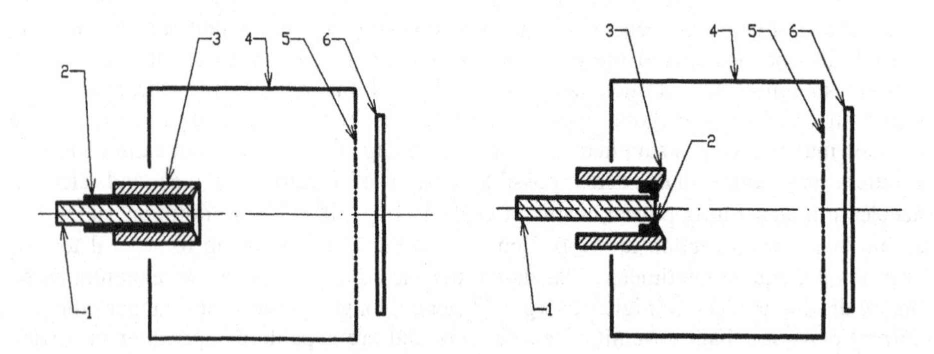

Figure 1. Schematic drawings of plasma cathode electron gun and general trigger systems. 1 – cathode, 2 – insulator, 3 – trigger electrode, 4 – hollow anode, 5 – emitting grid, 6 – collector.

oscillograms of the electron beam current obtained in the "vacuum" mode of operation of the discharge.

Increasing the gas pressure in the discharge gap causes part of the electrons emitted from the boundary of a cathode flare to ionize the gas. As a result, at a pressure $P > P_{cr}$ a different mechanism of the plasma formation occurs and, in a number of cases, it is faster than the mere propagation of the cathode flare. This results in a decrease in discharge initiation time [4]. The regime given is most suitable for a plasma electron emitter. In this mode, we have managed to obtain an electron beam with a high stability of the current waveform and amplitude (Fig.2b) and of the current density distribution over the beam cross section.

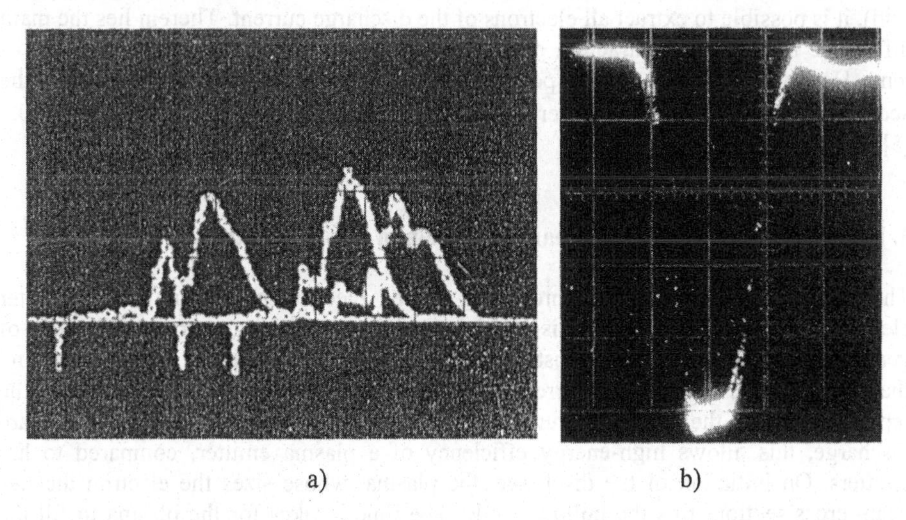

Figure 2. Electron beam current waveforms. For left figure - vertical scale: 125 A/div, horizontal scale: 2,5 μs/div; for right figure – vertical scale: 200 A/div, 1 μs/div.

Electrons are extracted from the plasma into the acceleration gap through the holes made at one of the surfaces of the hollow anode and covered with fine metal grid. At high emissivity of the cathode spot and high discharge current, the plasma produced in the hollow anode exhibits a high density and, with a certain design of the discharge system, a reasonably uniform plasma density distribution. Owing to the large anode area, the plasma has a potential φ_{Π} positive with respect to the anode potential φ_a. Therefore, electrons come to the anode walls through a potential barrier $\Delta\varphi = \varphi p - \varphi_a$ with the current density given by:

$$j=en(kT_e/2\pi m)^{1/2}\exp[-e\Delta\varphi/kT_e], \quad (1)$$

where e is the electron charge, k is Boltzmann's constant, n is the plasma density, T_e is the plasma electron temperature. The width of the potential barrier is determined from the equality of the Bohm ion current to the ion current in the layer following from the "3/2-power" law:

$$l_c=2/3(\varepsilon_0/0,4n)^{1/2}(2/ekT_e)^{1/4}\Delta\varphi^{3/4}. \quad (2)$$

Electrons are extracted and accelerated by the voltage applied between the hollow anode *4* and collector *6*. The efficiency of a plasma emitter is governed by the fact that in certain conditions the electron component of the discharge current can be extracted almost completely and the extraction efficiency $\alpha = I_e/I_d \cong 1$ where I_e is the emission current and I_d is the discharge current can be obtained. With no accelerating voltage, if the layer width l_s are greater than the half-width of the grid mesh $l_m/2$, all discharge electrons arrive at the walls of the hollow anode and at the emission grid wires. As the accelerating voltage is increased, the potential barrier in the grid holes becomes lower and more and more electrons come into the acceleration gap, thus increasing the plasma potential with respect to the anode potential. With some relation between the grid mesh size, the plasma density, and the accelerating voltage (the electric filed strength near the grid), it is possible to extract all electrons of the discharge current. Therein lies the main difference between the extraction of electrons from the plasma and the extraction of ions. However, to avoid strong perturbations in the plasma and breakdown of the acceleration gap, the plasma emitter is operated in the standard mode with $\alpha \leq 0.8 - 0.9$ [5].

3. Methods of pulsed electron beam formation in plasma emitters

The most popular method for producing pulsed electron beams in plasma-emitter electron sources and accelerators is "modulation" of the discharge current. This way of generating pulsed beams is the most simple and widely employed one. The duration and the repetition rate of the beam current pulses are determined by the duration and by the repetition rate of the discharge current pulses. At a low operating voltage of the arc discharge, this allows high-energy efficiency of a plasma emitter, compared to hot emitters. On initiation of the discharge, the plasma, whose sizes the electron dictates beam cross section, fills the hollow anode. The time it takes for the plasma to fill the cavity or the time it takes for it to be produced, which depends on the cavity volume, on the operating pressure of the plasma-forming gas, and on a number of other factors, is at best several microseconds [6]. The plasma formation is responsible for the emission current rise time and the plasma decay in the cavity for the decrease in electron current. Thus, the both processes dictate the pulsed electron beam time and frequency parameters. The first process limits the risetime and the minimum duration of the electron beam current pulses, while the second one limits their maximum repetition rate. The amplitude of the emission current or its density is mainly limited by the electric strength of the acceleration gap.

One of the methods for producing electron beams of pulse duration much shorter than the time it takes for the emitting plasma to be formed is that with emission grid control [7]. An electron emitter, which employs the given method, is shown schematically in Fig. 3. The emitter has three electrodes whose holes are covered by fine grids of high transparency. The emission boundary of the plasma generated inside cavity *2* is formed by grid electrode *3*. Grid electrode *4* is used to separate electrons and ions arriving from the plasma. The ions are reflected from a potential barrier occurring where a positive potential is applied to the grid *4,* and the electrons readily penetrate through the grid to get to the space between the this grid and control grid *5*. A dc voltage is applied to the

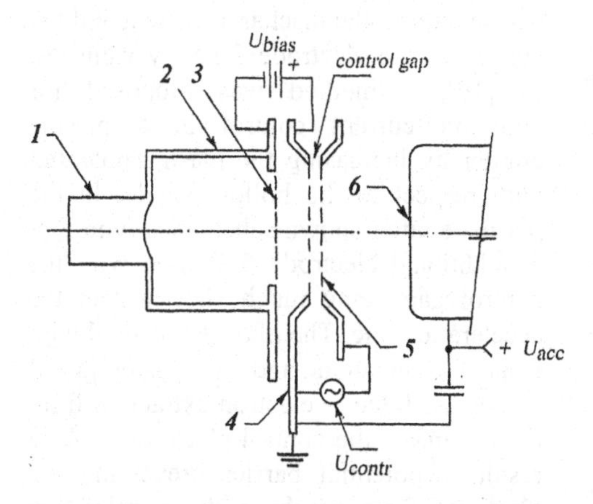

Figure 3. Schematic drawing of plasma cathode electron gun [6]. 1 – plasma generator, 2 – hollow anode, 3 – emitting grid, 4 – cathode grid, 5 – control grid, 6 – collector. U_{cont} – control signal, U_{bias} – negative bias, U_{acc} – accelerating voltage.

collector. By varying the potential of the electrode *5* with respect to that of the electrode *4*, it is possible to vary the amplitude of the emission current pulse and, with a definite amplitude and form of the control voltage, to vary the duration of the emission current pulse. Actually, in this emitter the well-known circuit of a triode with a hot cathode (all electrodes *1*, *2*, *3*, and *4*) and with a control grid is realized. The emitter at issue possesses reasonably good time and frequency characteristics, which depend exclusively on the electron transit time in the control and acceleration gaps.

However the use of two additional grid electrodes leads to a appreciable decrease in the electron extraction efficiency of the emitter due to the fact that part of the electrons is trapped by the grids. Therefore, in a number of cases where these losses are inadmissible (it is particularly related to electron beams with an amplitude of hundreds of amperes and higher), a plasma emitter with grid control presented in Fig. 4 can be used [8]. In the circuit proposed, emission grid *3* is insulated from cavity *2*. A negative voltage preventing the electrons from reaching the acceleration gap and a positive control voltage are applied to the grid *3*. Varying the positive control voltage can vary the emission current pulse amplitude and duration. Unlike the foregoing circuit, the time characteristics of the emission current in this system depend not only on the electron component of the plasma, but also on the ion one. Under certain conditions [9], the emission current risetime is no longer than a few nanoseconds and is determined by the plasma potential relaxation time, as compared to a hollow anode.

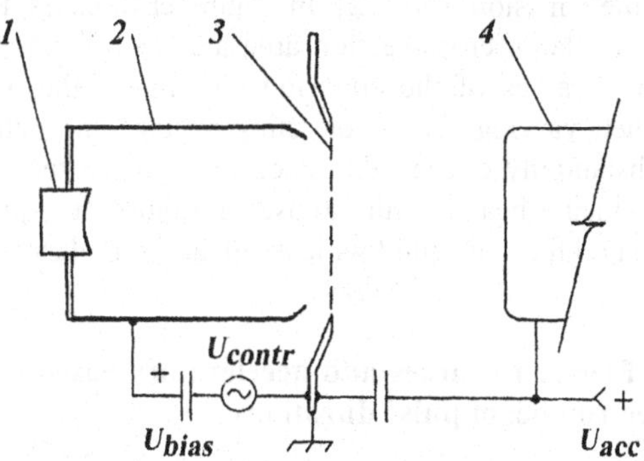

Figure 4. Schematic drawing of plasma cathode electron gun [7]. 1 – plasma generator, 2 – hollow anode, 3 – control grid, 4 – collector. U_{cont} – control signal, U_{bias} – negative bias, U_{acc} – accelerating voltage.

The plasma filling the cavity *2* is known to have a positive potential with respect to the anode. When electrons are extracted from the grid electrode *3*, the potential increases, thus decreasing the electron current in the circuit of the hollow anode. Under certain experimental conditions dependent on the grid mesh size and on the near-wall layer width, a mode can be realized with a high (up to 100%) electron extraction efficiency from the hollow anode where the emission

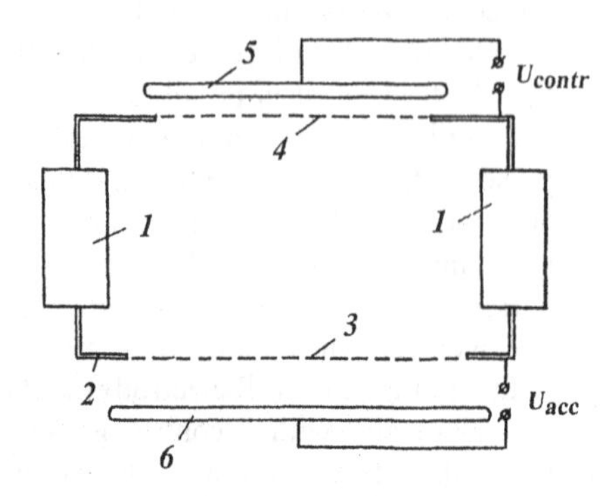

Figure 5. Schematic drawing of plasma cathode electron gun with additional control electrode [9].1 – plasma generators, 2 – hollow anode, 3 – emitting grid, 4 – additional grid, 5 – control electrode, 6 – collector.

current equals the discharge current and the current of the electrode *2* nearly vanishes. In [10], a method was proposed for emission current control in a plasma emitter by increasing the plasma potential with respect to the hollow anode. In the plasma emitter shown schematically in Fig. 5, additional electrode 5 is used with the control gap being much shorter than the acceleration one. The plasma potential with respect to the anode and to emission grid 3 is increased due to electron extraction from the plasma to the control electrode 5. As a result, a potential barrier preventing the electrons from reaching the acceleration gap occurs near the grid as well as near the walls of the hollow anode. A deep (up to 100%) electron beam current modulation with current pulse time parameters approaching those of the emitter with grid control (Fig. 4) can be accomplished by the proper choice of the grid mesh size, control gap length, and voltage applied to it A similar method based on the use of the plasma potential increase phenomenon in electron extraction to form and modulate the ion current was realized in a system of immersion ion implantation [11].

Along with the above methods for producing pulsed electron beams in plasma emitters with dc accelerating voltage, nanosecond and microsecond pulsed electron beams can be generated with the use of pulsed accelerating voltage. With no accelerating voltage, plasma penetrates into the acceleration gap and fills it. Once an accelerating voltage is applied, a mode close to short-circuiting occurs during the risetime of the voltage pulse. As the plasma decays, two modes of operation are possible: normal with voltage recovery across the acceleration gap or breakdown with the formation of a cathode spot at the emission grid [12]. In a number of cases, the need is generated in electron beams with a large cross-section area and with nanosecond pulse risetimes at comparatively small values of the emission current (of the order of several hundreds of amperes). When varying the accelerating voltage in such systems, the capacitive current may substantially exceed the electron beam current. In such a case, the given method for producing beams with nanosecond pulse durations turns out to be energy-useless. This method has not found wide application in plasma-emitter electron sources.

4. Electron sources and accelerators based on plasma emitters with microsecond electron beam pulse durations

Figure 6 shows schematically the design of an accelerator generating an electron beam of cross-section area 15×60 cm^2 [12]. This accelerator was used to produce laser radiation in inert gases. The plasma emitter is common cylindrical anode *1* with two cathode units of the discharge system mounted on its faces. The plasma-forming gas,

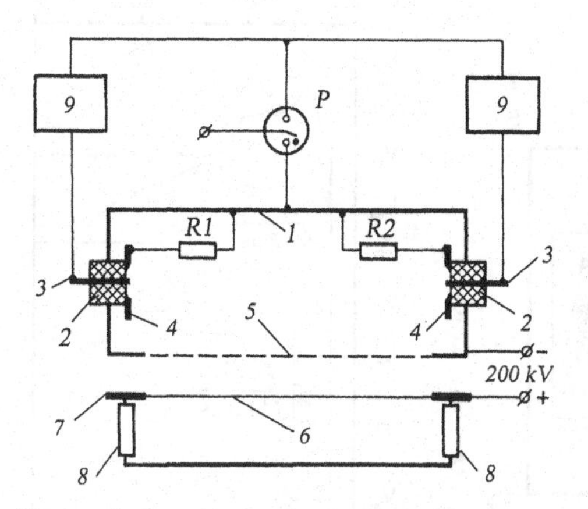

Figure 6. Electron accelerator with rectangular beam cross section [11]. 1 – hollow anode, 2 – insulator, 3 – cathodes, 4 – trigger electrodes, 5 – emitting grid, 6 – Al foil, 7 – support electrode, 8 – laser mirrors, 9 – pulse forming lines, P – gas switcher, R1, R2 – resistors.

mainly nitrogen or argon, is supplied to the discharge gap (into the hollow anode) up to an operating pressure of 2×10^{-2} Pa. On application of voltage, a surface discharge of dielectric *2* is initiated between electrodes *6, 7*. The intermediate electrode is connected to the anode through a resistor of resistance R=75 Ω. The current-limited surface discharge generates low-density plasma in the hollow anode, initiating an arc discharge between the cathode and the hollow anode. Such a system of connection of the trigger electrode makes it possible to do away with an additional power supply. On the side surface of the hollow anode, there is an emission window covered with a grid of mesh size 0.4×0.4 mm. The window dimensions determine the beam cross section. Electrons are extracted and accelerated by a dc voltage applied between the hollow anode and the vacuum chamber. When the beam is extracted through a thin foil into the atmosphere or into a high-pressure gas, the dc voltage makes it possible to minimize the electron losses and to produce a nearly monoenergetic beam owing to the absence of electrons which would be accelerated during the risetime and falltime of a high-voltage pulse.

For the electron beam pumping of lasers, an important characteristic is the uniformity of the current density distribution over the beam cross section that determines both the efficiency of utilization of the beam and the probability of breakdown of the working gas mixture of the laser at a local inhomogeneity of the beam. The method of improving the current density distribution is based on the dependence of the electron emission current on the proportion between the grid mesh size and the width of the negative potential fall layer near the anode wall. In the general case, the electron emission current into the acceleration gap can be presented in the form:

$$I=jS_e-jS_l[1-\exp(-e\Delta\Phi/kT_e)], \tag{3}$$

where S_e is the emission area, S_l is the total area of the layers near the grid wires. To make the current density more uniform, a grid of variable transparency is placed in the hollow anode in the region of elevated plasma density. Another way of improving the current density distribution lies in variation of the inclination angle γ (Fig, 7) of the cathode assembly with respect to the axis of the hollow anode.

With an accelerating voltage of 150-200 kV, a pulse repetition rate of 50 Hz, and a pulse duration of 30 μs, a beam of current up to 50-80 A was generated. The electron

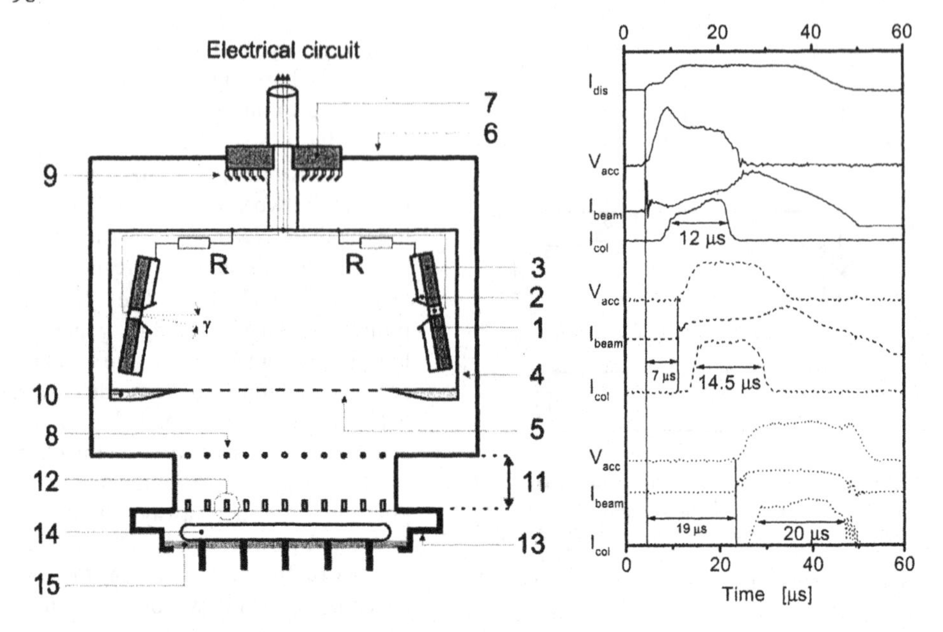

Figure 7. Schematic drawing of the gun [11]: 1 - discharge cathode, 2 – intermediate electrode, 3 - insulator, 4 - hollow anode, 5 - extraction grid, 6 - vacuum chamber, 7 - high-voltage insulator, 8 - grid anode, 9 - voltage gradient rings, 10 - field shaping electrode, 11 - drift section, 12 - Ti foil window and support structure, 13 – laser chamber, 14 - discharge electrode and 15 - insulator.

Figure 8. Waves of the discharge current I_{dis}, the accelerating voltage V_{acc}, beam current I_{beam} and collector current I_{col}. When the accelerating voltage and discharge current start simultaneously, the beam current rises only slowly (solid curves). With sufficient delay the beam current rises immediately (dashed curves). Finally, by adjusting the falling slopes of the discharge current and accelerating voltage it is possible to eliminate the high-voltage discharge at the end of the pulses (dotted curves).

beam was extracted into the atmosphere through a 40-μm titanium foil. With a 70% geometric transparency of the support grid cooled with water at the periphery, 50% of the beam current is extracted from the acceleration gap. In the single-pulse operation, the accelerator produces a beam of current up to 1000 A at a pulse duration of 5-10 μs, of current up to 400 A at a pulse duration of 15 μs, and of current up to 200 A at a pulse duration of up to 30 μs. With a pulse duration of up to 100 μs, the beam current amplitude is 25-90 A.

An electron accelerator where the current pulse duration depends on the accelerating voltage pulse duration, rather than on the discharge initiation and operation times is described elsewhere [13]. The accelerator is schematically shown in Fig. 7 and oscillograms of the discharge current, accelerating voltage, beam current, and collector current downstream of the foil is presented in Fig. 8. With no accelerating voltage, the discharge plasma from the hollow anode gets into the acceleration gap and fills it. On application of accelerating voltage, the total voltage is applied to a narrow region adjacent to the emission grid. This results in an increase in electric field strength near the grid electrode and, on approaching some value of the field strength, in emission centers at the grid surface. Near the emission grid, dense plasma is formed which, expanding toward the outlet grid, causes shortcircuiting of the gap. This is one of the plausible mechanisms for the breakdown of the acceleration gap. With constant accelerating voltage and risetime, the penetrating plasma density determines the size of this region. The plasma density depends on the value of the discharge current in the discharge system, on the grid mesh size, and on the time delay between the moments the discharge is initiated and the moment the accelerating voltage is applied. The beam and collector current pulse form and the voltage pulse form across the acceleration gap also depend on this delay time, since the pulse voltage generator has a finite internal resistance (Fig. 8). By properly choosing the foregoing parameters, we have managed to realize the stable operation of the accelerator in the following mode: beam current 270 A, accelerating voltage amplitude 200 kV, collector current pulse duration (), and 65% beam losses in the extraction grid and in the foil.

5. Electron sources and accelerators based plasma emitters with nanosecond beam current pulse duration

The time it takes for the plasma to be formed in the anode cavity depends to a great extent on the pressure of the background or plasma-forming gas and, at very low pressures, it is determined by the plasma velocity, which is nearly equal to 10^6 cm/s. In this connection, in electron emitters, which are employed to produce electron beams with a cross section of tens of centimeters, the emission current risetime can amount to tens of microseconds. The grid control method allows generation of electron beams

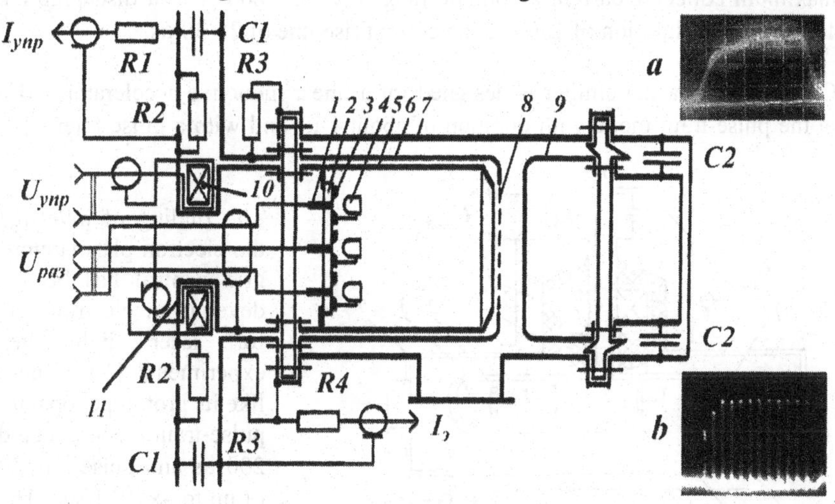

Figure 9. Plasma electron emitter with grid control: 1 – cathode, 2 – insulator, 3 – trigger electrode, 4 – redistributing electrode, 5 – hollow anode, 6 – control electrode, 7 – vacuum chamber, 8 – emitting grid, 9 – collector, 10 - pulse transformer, and 11 – second winding of the pulse transformer. Oscillograms: waveform of emission current pulse (upper trace, vertical scan is 210 A/div, sweep is 20 ns/div) and 12-pulse burst with 33 μs delay (lower trace).

with specified time characteristics (risetime and falltime), which are much smaller than the plasma formation time.

Figure 9 shows schematically the design of an electron plasma emitter with grid control [14]. In this emitter, as in those described above, the emitting plasma is formed due to an arc discharge operating between cathode 1 and hollow anode 5 both in the vapors of the cathode material and in the gas fed into the anode cavity through a controllable leak. The most optimum pressure of the supplied gas lies in the range from 10^{-2} Pa to 3×10^{-2} Pa. To increase the lifetime and improve the uniformity, seven triggering systems with seven cathodes and seven trigger electrodes, accordingly, were employed. Each of the cathodes was connected to an individual artificial pulse-forming line. All trigger electrodes were connected through 50-100 Ω resistors to the hollow anode. A dc negative voltage of up to 100 V was applied to grid emission electrode 8 separated from the hollow anode, which, in fact, prevented the plasma electrons from reaching the acceleration gap between the emission grid and collector 9. A dc accelerating voltage was applied between the emission electrode and the collector. Once the hollow anode was filled with plasma, positive pulses from a nanosecond pulsed generator were applied to the emission grid to decrease the potential barrier near the grid and to give rise to the emission current. The pulse transformer of the generator and the plasma emitter were integrated to achieve better time parameters. An oscillogram of 10 superimposed electron beam current pulses is shown in Fig. 9 in the upper right-hand corner. It can be seen from the oscillogram that the pulse and amplitude waveforms nearly replicate each other and the plasma emitter features a fairly high stability. The maximum collector current amplitude ranged to 700÷800 A at a discharge current of 1300 A, pulse duration of 100 μs, and current risetime of 25-30 μs.

Our electron plasma emitter is designed for a linear inductive accelerator and operates in the pulse-train mode with duration of about 1 μs and with a pulse interval of 30-33 μs.

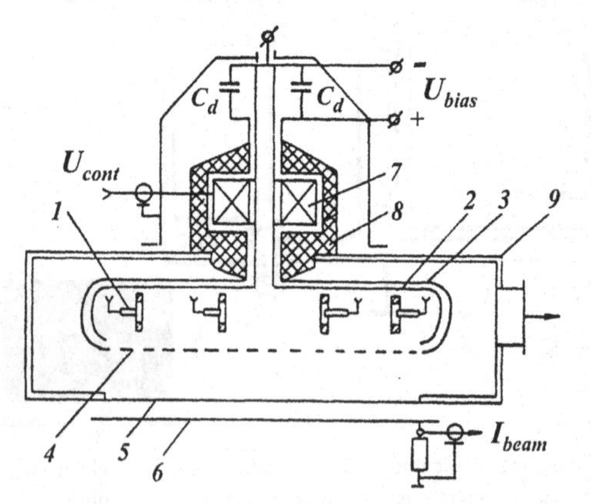

Figure 10. Schematic drawing of an accelerator with a high pulse repetition rate [19]. 1 – cathode unit, 2 – hollow anode, 3 – outer cylindrical electrode, 4 – emitting control grid, 5 – Ti foil with support structure, 6 – collector, 7 – pulse transformer, 8 – feedthrough insulator, 9 vacuum chamber.

The finding of investigation of the electron plasma emitter with grid control has been used to develop an electron accelerator for laser light generation experiments [15]. The emitter, like its prototype, operates in the pulse-train mode with a duration 200 μs and pulse repetition rate of up to 4×10^5 1/sec. The circuit of the accelerator is shown in Fig. 10. In vacuum chamber 1, with the help of bushing insulator 8 and a current lead, emitter 3 is mounted. Plasma is produced by four cathode units 1 with arc discharges initiated by application of voltage from pulse-forming lines connected to the output stage of a voltage pulse generator. The pulsed discharge current of duration 200 μs is controlled in the range from 0.3 to 1 kA. A negative bias voltage U_{bias} of up to 400 V is applied between hollow anode 2 and emission grid 4. Positive bell-shaped control voltage pulses of 170 ns FWHM and 3 kV amplitude are applied from the generator trough pulse transformer 7 mounted in the insulator body. With an accelerating voltage of 160 kV, a discharge current of 400 A, a 70% geometric transparency of the support grid, and a 18-μm thickness of the titanium foil, an electron beam of current 100 A and cross-section area 3x70 cm^2 is extracted. Estimates show that in an accelerator with grid-controlled electron emission the pulse repetition rate can be increased to several megahertz.

6. Electron source based on a plasma emitter operating in the "vacuum" mode

One of the promising applications of an electron plasma emitter is its use in an ion source based on a vacuum arc operating in the vapors of the cathode material (MEVVA). This provides an increase in the average charge of the ion beam. The injection of a low-energy electron beam into the plasma produced in the ion source causes an increase in ion charge [16]. In the forgoing electron accelerators, the discharge in the hollow anode of the emitter operated at a pressure of 3×10^{-2} Pa, which is in excess of the critical pressure p_{cr} for the given geometric dimensions. At such a pressure, a discharge reminiscent of a gas discharge is initiated in the discharge gap and the gas discharge plasma is formed. Experiments performed to study the composition of an ion beam extracted from an ion source with a discharge gap geometry similar to the described above have shown [17] that at a pressure higher than 2×10^{-2} Pa the metal ion fraction (the cathode material, e.g. Al) decreases to 20% and lower. As this takes place, 80% of the ion beam fall at ions of the gas fed in the discharge gap. Such a relatively

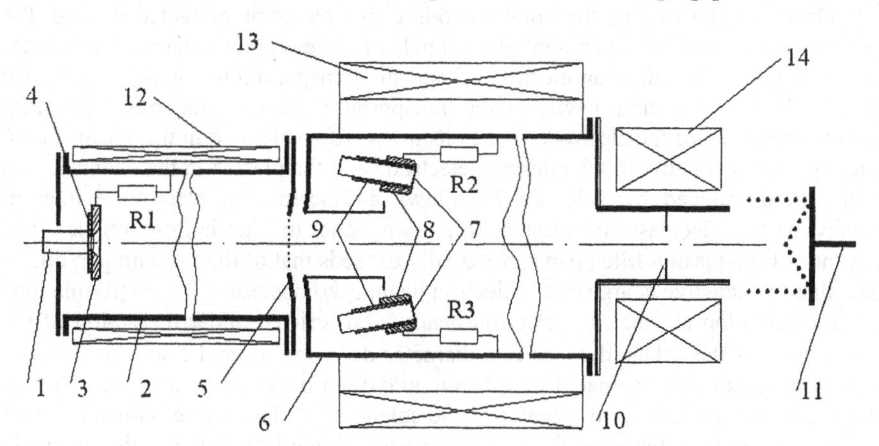

Figure 11. Schematic drawing of electron gun [14]. 1, 7 – cathodes, 2 – hollow anode, 3, 8 – trigger electrodes, 4, 9 – insulators, 5 – emitting electrode, 6 – transport channel, 10 – diaphragm, 11 – collector (Faraday cup or plate), 12, 13, 14 – coils.

high pressure in an electron source stabilizes the emission current, yet being entirely unsuited for metal ion sources. In [18], it has been shown that the average charge of metal ions decreases substantially, even at a pressure higher than 6×10^{-4} Pa.

To be used in an ion source with an electron beam (E-MEVVA), a plasma emitter has been designed based on an arc discharge with a cathode spot where the emission current density is 20÷40 A/cm^2. The emitter operates in the conditions of high vacuum ($p\leq2\times10^{-4}$ Pa) and features high extraction efficiency [19]. An experimental prototype of a plasma-emitter electron gun, on which the design of the electron source has been developed and optimized, is shown schematically in Fig. 11. The prototype consists of two main units: a plasma emitter and an electron beam transportation and focusing system. The plasma emitter is formed by cold cathode *1*, hollow anode *2*, and trigger electrode *3*. At one of the hollow anode faces, there is emission window *5* of diameter 1.5 cm covered with a fine grid. The plasma is in a longitudinal magnetic field of 0.1÷0.2 T of solenoid *12*. The beam is transported and focused in channel *6* inside which plasma generator guns are placed. The extraction electrode is made in the form of a disc with a hole of diameter 1.8 cm having no grid cover. To compensate the defocusing action of the electric fields in the hole of the extraction electrode, the transportation channel and the volume inside the hole were preliminary filled with plasma produced by the plasma generators and were immersed into the magnetic field of solenoid *13*. At the outlet of the transportation channel there is focusing coil *14*. The maximum values of the magnetic field along the coil axis reach 1 T.

The electron gum operates as follows. Voltage is first applied to the solenoids and then, after a time longer than that it takes of the magnetic field to penetrate into the hollow anode and into the transportation channel, to the electrodes of the plasma generators. Once the transportation channel is filled with plasma, an arc discharge is initiated inside the plasma emitter. The emitter discharge current is controlled in the range from 50 to 150 A. Once the plasma fills the hollow anode *2*, the electrons extracted through the grid meshes come into the transportation channel. An accelerating voltage of 10-20 kV is applied between the emission electrode *5* and the transportation channel *6*. Because the plasma filing the internal cavity of the transportation channel acquires a potential close to the potential of the electrode *6*, the electrons are accelerated in the space charge double layer formed between the emission electrode and the plasma in the channel. The magnetic field induced by solenoid *12* allows an increase in electron extraction efficiency and a decrease in plasma formation time in the hollow anode. The purveyance of the plasma-filled gap substantially exceeds that of the vacuum gap due to the fact that the negative charge of the electron beam is compensated by the plasma ions. In the transportation channel, the electron beam is constricted under the action of the magnetic field of the solenoid *13* and its diameter decreases from 1.5 am to ≤1.1 cm. The beam diameter was estimated by a beam autograph obtained on a plane collector located along the gun axis at the center of the solenoid *13*. The further constriction of the electron beam to a diameter close to 8 mm was accomplished due to the magnetic field induced by the coil *14*. Downstream of the focusing foil, the beam diameter increased to nearly 2 cm at a distance of 10 cm from the coil.

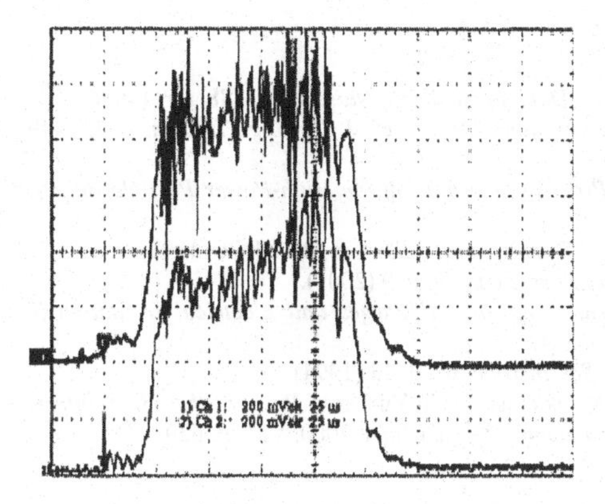

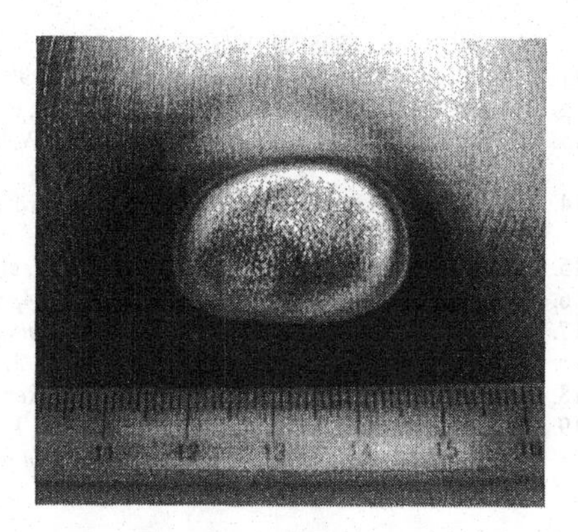

Figure 12. Left figure is waveforms of collector (upper trace) and emission currents (bottom trace). Vertical scale: 20 A/div, horizontal scale: 25 μs/div. Right figure is electron beam print (about 50 J/cm^2).

In the course of experiments, an electron beam with a current of up to 80 A has been obtained at a pulse duration of up to 100 μs and accelerating voltage of 20 kV. Oscillograms of the emission and collector currents and also the beam autograph on the collector are shown in Fig. 12. The current density at the collector was 25-30 A/cm^2, while the current density in the hole of diaphragm 10 reached 160 A/cm^2.

The collector plasma and the vapors of the collector material, which occurred on the bombardment of the collector by the electron beam, did not exert any profound effect on the electric strength of the acceleration gap. This is due to the fact that the action of electrons penetrating through the hole in the focusing coil to the transportation channel and then to the acceleration gap is small. Such a circuit of an electron source holds promise in designing technological systems, which use the electron beam for pulse-induced melting of the material surface to modify its properties.

A slightly modified version of the electron source was used experimentally to increase the average charge in ion sources of the MEEVA type [16].

References:

1. Gilmour A. S., Jr., Lockwood D. L. *Proc. IEEE.* **60**, 977 (1972).
2. Koval N.N., Kreindel Yu.E., Mesyats G.A., et al. *Pisma Sov. Tekh. Fiz,* **9**, 568 (1983).
3. Koval N.N., Kreindel Yu.E., Tolkachev V.S., Schanin P.M. *IEEE Trans. on Electr. Insul.* **20**, 735 (1985).
4. Kozyrev A.V., Korolev Yu. D., Shemyakin I.A. *Rus. Phys. J.* **37**, (1994).
5. Gavrilov N.V., Gushenets V.I., Koval N.N., Oks E.M., Tolkachev V.S., Schanin P.M. In book: *"Plasma emitter of a charged particle".* Ekaterinburg, Nauka, 1993.
6. Gushenets V.I., Koval' N.N., Schanin P.M. *Proc. of the VIth All-Union Symp. on High Current Electronics,* Tomsk, USSR, 1986. V. 2. p.112.
7. Humphries S., Jr., Coffey S., Savage M., et al. *J. Appl. Phys.* **57**, 709 (1985).
8. Gushenets V.I.; Koval' N.N.; Kreindel' Yu.E.; Schanin P.M. *Zhur. Tekh. Fiz,* **57**, 2264 (1987).
9. Galansky V.L., Gushenets V.I., Oks. E.M. *Proc. of the VIIth All-Union Symp. on High Current Electronics,* Tomsk, USSR, 1988, V. 2, p. 174.
10. Gushenets V.I., Kreindel' Yu.E., *Method of production of a pulse electron beams.* Author's Cert. No.

1400462, MKI NO5 N1/00/ (April 14, 1986).
11. Gunzel R. *J. Vac. Sci. Technol. B*, **17**, 895 (1999).
12. Vintizenko L.G., Gushenets V.I., Koval' N.N., et al. *Doklady Akademii Nauk SSSR,* **288,** 609 (1986).
13. Gielkens S.W.A., Peters P.J.M., Witteman W.J. Gushenets V. I., et al. *Rev. Sci. Instrum.*, **67**, 2449 (1996).
14. Bugaev S.P., Gushenets V.I. and Schanin P.M *Proc. IX Intern. Conf. on High-Power Particle Beams,* Washington, USA. 1992. V. 2, p. 1099.
15. Gushenets V.I.; Koval' N.N.; Kuznetsov D.L., et al. *Pisma. Zhur. Tekh. Fiz.*, **23-24,** 26 (1991).
16. Bugaev A., Gushenets V., Yushkov G., et al. *Appl. Phys. Lett.*, **79**, 919 (2001).
17. Oks E., Yushkov G., *Proc. of the XVIIth Intern. Symp. on Discharges and Electrical Insulation in Vacuum,* Berkeley, California, USA, 1996, V. 2, p. 584.
18. Spadtke P., Emig H., Wolf B.H., and Oks E. *Rev. Sci. Instrum.* **65,** 3113 (1994).
19. A.S.Bugaev, V.I.Gushenets, Yu.A. Khuzeev, E.M.Oks, and G.Yu.Yushkov. *Proc. of the XIXth Intern. Symp. on Discharges and Electrical Insulation in Vacuum,* Xi'an, China, 2000, V. 2, p. 629.

EMISSION METHODS OF EXPERIMENTAL INVESTIGATIONS OF ION VELOCITIES IN VACUUM ARC PLASMAS

A.S. Bugaev, V.I. Gushenets, A.G. Nikolaev, E.M. Oks,
G.Yu. Yushkov, A. Anders*, I.G. Brown*
High Current Electronics Institute Russian Academy of Sciences
4 Academichesky ave., Tomsk
Russia 634055
**Lawrence Berkeley National Laboratory, University of California,*
1 Cyclotron Road, Mailstop 53, Berkeley, California
USA, 94720

Abstract. This paper is devoted to an investigation of the directional velocities of the ions generated in cathode spots of vacuum arc discharges. Using emission methods of studying the processes in a vacuum arc discharge, which involve the determination of the parameters and characteristics of the discharge plasma by analyzing the ion current extracted from the plasma and the ion charge states, the velocities of ions have been determined for the majority of cathode materials available in the periodic table. Is has been shown that at a low pressure of the residual gas in the discharge gap the directional velocities of the ions do not depend on the ion charge state. Comparison of the data obtained with calculated values allows the conclusion that the acceleration of ions in a vacuum arc occurs by the magnetohydrodynamic mechanism.

1. Introduction

Among the key experimental facts established in studying the vacuum arc discharge [1, 2], an important one is the existence of directional flows of ions emitted by cathode spots and moving toward the anode with an energy exceeding the operating voltage of the discharge [3, 4]. Now there is no consensus of opinion regarding the physical mechanism of the acceleration of ions, which is, first of all, due to the inconsistency of the available experimental data.

In spite of the fact that Davis and Miller [5] have found that the ion velocity increases in direct proportion with its charge, in some other experiments, for example, in the experiment of Tsuruta et al. [6], under certain conditions the measured velocities of ions practically did not depend on the ion charge state.

The emission methods developed by us to investigate the processes in a vacuum arc discharge [7, 8] involve the determination of plasma parameters and characteristics

E. Oks and I. Brown (eds.),
Emerging Applications of Vacuum-Arc-Produced Plasma, Ion and Electron Beams, 105–113.

based on the analysis of the ion current extracted from the plasma and its charge states. In this paper, the results of the determination of directional velocities of ions in a vacuum arc and their comparison with predicted values are given, allowing a conclusion about the mechanism of the acceleration of ions in a vacuum arc.

2. The emission method for investigating ion velocities

The basic methodical difficulties arising in studying the parameters of the plasma of a vacuum arc are related to the fact that the cathode spot occupies a random position and moves with a velocity of $\sim 10^4$ cm/s over the cathode surface, it is small in size ($< 10^{-1}$ cm), and the plasma density in the region adjacent to the cathode spot is over 10^{18} cm^3. These features of the cathode spot as a physical object impede its experimental investigation. In this connection, methods of corpuscular diagnostics, in particular the emission method [7, 8], seem to offer, perhaps, a solely probable approach for studying experimentally the processes in a cathode spot. The idea of the method is to study of the response of the charge state distribution of the ions extracted from the emission boundary located far away from the cathode spots of a vacuum arc to the action of which the discharge is subjected. Such an action may be the abrupt change in vacuum arc current resulting in the death of existing cathode spots or birth of new ones.

The electrode arrangement of the experimental setup is given in Fig. 1. The vacuum arc discharge was initiated in a discharge gap between cathode *1* and anode *3*. The duration and current of the discharge pulse were 200–500 μs and 100–500 s, respectively. During the operation of the arc, cathode material plasma *4* emitted by cathode spots filled the cavity of the anode. There were emission holes in the face surface of the anode designed for extraction of ions from the plasma. The extracted ions were accelerated in multiaperture acceleration/deceleration system *5* by dc voltage of 10–30 kV. Subsequently the charge state constitution of ion flow *6* was analyzed by a time-of-flight mass spectrometer [9, 10].

In over 150 μs from the initiation of the vacuum arc, when all its basic parameters can be considered steady-stated, an additional current pulse was applied to the discharge gap or the power supply of the discharge was closed by a fast switch [11]. In the first

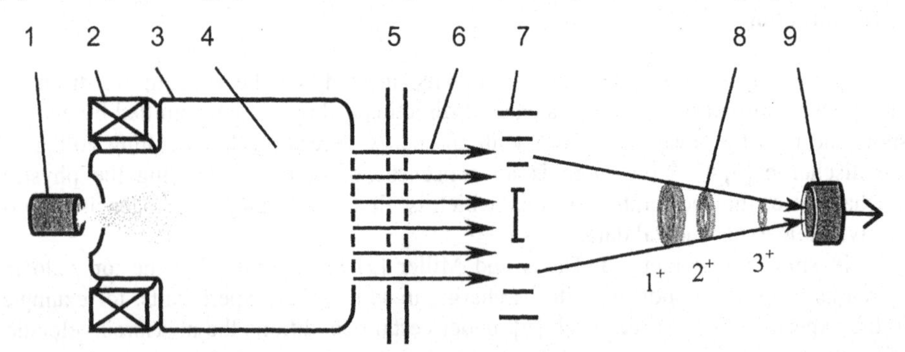

Figure 1. The electrode arrangement of the setup for measuring ion velocities: *1* – cathode; *2* – solenoid; *3* – anode; *4* – vacuum arc plasma flow; *5* – acceleration electrodes; *6* – ion beam; 7 – spectrometer gate; *8* – ions of various charge state; *9* – Faraday cup.

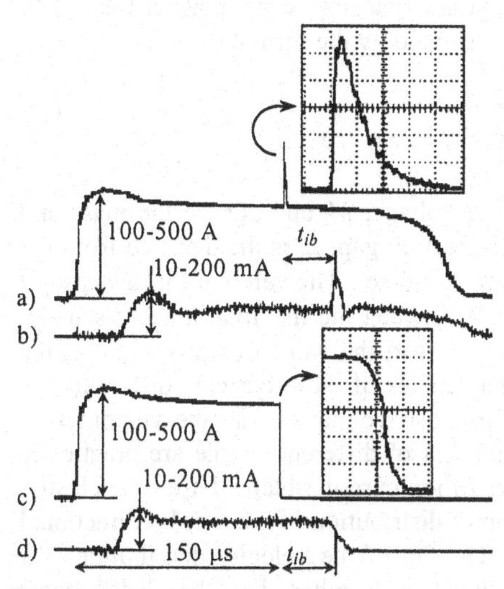

Figure 2. Oscillograms of the vacuum arc (*a*, *c*) and ion beam currents (*b*, *d*) in cases of a burst (*a*, *b*) and cutoff (*c*, *d*) of a current of an arc. Oscillograms of the arc current recorded on application of a current burst and during current cutoff are given in the insets: 1 μs/div, 25 A/div.

case, the arc current increased by 50–150 A within 2–4 μs, while in the second one it decreased to zero in about 2 μs, leading to extinction of the discharge (Fig. 2). In what follows, when describing the mentioned changes in arc current, the term's "current burst" and "current cutoff" will be used.

Attention is drawn to the fact that both in the case of an arc current burst and in the case of its cutoff the response of the ion beam current to the mentioned perturbations of the arc current was observed after some time interval t_{ib}. The measured values of t_{ib} depend on the cathode material being 8 μs for the lightest material (C) and 40 μs for the heaviest one (Bi). At the same time, the travel times for the ions of these elements in the acceleration gap and in the drift space were less than 1 and 3 μs, respectively. It is obvious that this considerable difference in t_{ib} is due to the motion of the ions in the plasma from the cathode region, in which they have been generated, to the emission grid. By the change in current of each ion species it was possible to determine the time after the action on the vacuum arc in which the change in the emission current extracted from the discharge plasma occurred and thus find the velocity of motion of ions from the vacuum arc cathode spot to the emission surface.

3. Influence of the charge of ions on their directional velocity in plasma

Typical dependencies of the current on the time after the application of an additional current burst to the discharge gap for ions of different charge state obtained for the case of a vacuum arc with a magnesium cathode are given in Fig. 3. The same figure presents similar dependencies for the case the arc current was cut off. It should be noted that when a current burst is applied to the discharge gap, the currents of ions of each charge state reach a maximum at the same time, while in the case the discharge current is cut off, the fall in current occurs identically for variously charged ions. These features were observed for all cathode materials used in the experiment, namely, Mg, Al, Ti, Zr, Cu, Pb, and Bi.

Both in the case of an arc current burst and in the case of its cutoff the response of the ion current shows up in a certain time *t*. Since the basic processes of ionization of the cathode material in a vacuum arc occur near cathode spots, at distances which are not over 1 mm from the cathode surface and much smaller than the cathode–anode separation $L_{c\text{-}a}$, and are followed only by adiabatic expansion of the plasma flare [12], the velocity of ions in the plasma, v_i, can be determined by the formula:

$$v_{\mathrm{i}} = \frac{L_{\text{c-a}}}{t - \left(L_3 + 2d_{\text{accel}}\right) \cdot \sqrt{M_{\mathrm{i}}/2QeU_{\text{accel}}}} \quad , \tag{1}$$

where V_{accel} is the amplitude of the accelerating voltage; M_i and eQ are the mass and charge of the ions; d_{accel} is the width of the acceleration gap; L_3 is the distance from the acceleration electrode to the gate of the mass spectrometer. The value of t in the case of an arc current burst was determined as the time between the maxima of the discharge and the ion currents, while in the case of current cutoff the value of t was taken as the interval between the maxima of the time dependencies of these currents differentiated with respect to time. The investigations performed have shown that the velocities of directional motion of ions of the same material, but of different charge are practically identical. Differentiating the time dependencies of ion current obtained after the closure of the discharge gap, we can obtain the function of distribution of the ions by directional velocities. Thus we obtain that not only the maxima of the velocity distributions for different ionic species practically coincide with each other, but the distributions themselves as well. A magnetic field created in the discharge gap increased the directional velocities of ions; however, the currents of ions of different charge for each cathode material varied proportionally.

The data obtained do not agree, first of all, with the experimental results of Tsuruta et al. [5] who measured the dependence of the velocity of ions, v_i, on their charge. It is obvious that this disagreement should be due to different experimental conditions. Essential differences were that in the experiment [5] the arc discharge operated during $t = 0.5$ s, while in the experiment under consideration t was shorter than 0.5 ms and that in [5] the discharge device was evacuated with an ion pump at a rate of only 20 l/s. The long duration of the discharge pulse and the rather small evacuation rate provided by the pump used in combination with the "falling" dependence of the evacuation rate quantity on pressure inherent in this type of pump could have the result that the pressure in the discharge system during the operation the vacuum arc was much higher than that reported in [5].

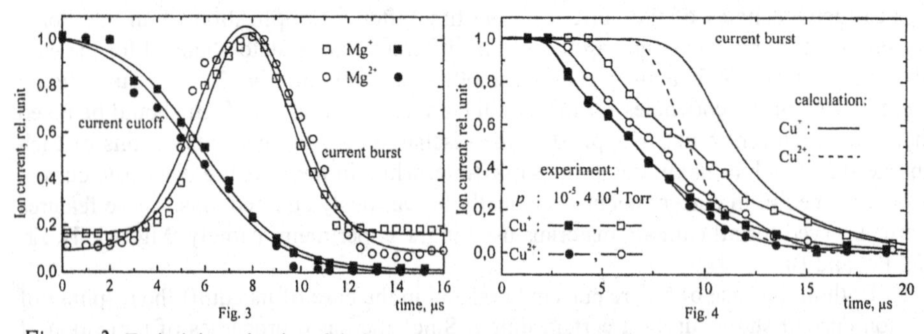

Figure 3. Typical dependencies of the current on the time after the application of an additional current burst to the discharge gap for ions of different charge state. Cathode – Mg.

Figure 4. Dependencies of the ion current on the time after the cutoff of the arc current for Cu^+ and Cu_2^+ ions.

To verify this hypothesis, measurements of the velocities of ions were carried out with forced supply of gas in the discharge gap. An increase in pressure p of the gas as it was supply into the discharge gap resulted in an increase in the time at which the response of the ion current to a perturbation of the arc. At the same time, the effect of an increase in p was different for different charge states. As an example, we present the dependencies of the ion current on the time after the cutoff of the arc current for Cu^+ and Cu_2^+ ions (Fig. 4) for the case where Ar was supplied into the discharge system. In the same figure, the dependencies calculated based on the data obtained by the authors of [5] are given. It can be seen that the tendency in changing the dependencies with increasing gas pressure is toward a disproportionate change in the velocity of singly and doubly charged ions and as the pressure is increased, the dependencies approach those that should be observed according to the data of [5]. Unfortunately, in the test discharge system, owing to the necessity of maintenance of the electrical strength of the acceleration gap, the peak pressure was limited. However, the disproportionate change in velocities of variously charged ions has been observed unambiguously for various cathode materials and various gases supplied into the discharge gap.

4. Directional velocities of ions for various cathode materials

Since, as has been well established, the directional velocities of various ionic species are practically identical for $p < 5\cdot10^{-5}$ Torr, the use of a time-of-flight spectrometer, intended for separation of the ionic component of a plasma by species, was not necessary for further investigation of directional velocities of ions. Detailed investigations of the dependence of the directional velocities of ions on the arc operating voltage were carried out with the use of the discharge system of the Mevva-5 ion source [13]. These experiments featured a slower modulation of the arc current pulse with characteristic oscillation time t » 50 μs and amplitude not exceeding 30 % of the pulse current and the use of a flat single Langmuir probe for measuring the ionic component of the plasma. The velocity of ions in the given experiments was estimated as $v_i = L_{c\text{-}p}/\Delta t$, where $L_{c\text{-}p}$ is the distance from the cathode to the probe surface and Δt is the time shift between the oscillations of the discharge and probe currents (Fig. 5). The measured velocities of ions for various cathode materials, including the majority of conductive materials of the periodic table, and the arc operating voltages are given in Table 1.

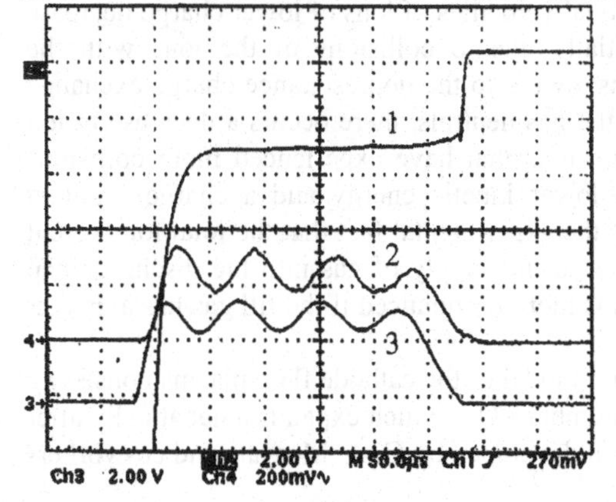

Figure 5. Oscillograms of the operating voltage (*1* – 20 V/div.), the ion current onto the probe (*2* – 0.1 A/div.) and the discharge current (*3* – 200 A/div.) for a vacuum arc with a tantalum cathode. Distance from the cathode to the probe: 24 cm. The shift between the oscillations of the arc current and the ion current onto the probe was 20 μs, which corresponds to v_i of tantalum ions, $1.2\cdot10^6$ cm/s.

Table 1. Measurements of directional ion velocities. V_d is the arc operation voltage, <Q> is the average ion charge, T_e is the plasma electron temperature [14], $I_d = 250$ A, $p < 7\cdot10^{-7}$ Torr.

Material	$v_i\cdot10^{-6}$, cm/s	V_d, V	<Q>	T_e, eV	Material	$v_i\cdot10^{-6}$, cm/s	V_d, V	<Q>	T_e, eV
Li	2.3	23.5	1	2	Cd	0.7	14.7	1.3	2.1
C	1.7	31	1	2	In	0.6	16	1.3	2.1
Mg	2	18.6	1.5	2.1	Sn	0.7	17.4	1.5	2.1
Al	1.5	22.6	1.7	3.1	Ba	0.8	16.5	2	2.3
Si	1.5	21	1.4	2	La	0.7	18.7	2.2	1.4
Ca	1.4	20.5	1.9	2.2	Ce	0.8	17.6	2.1	1.7
Sc	1.5	21.6	1,8	2.4	Pr	0.8	20.5	2.3	2.5
Ti	1.5	22.1	2	3.2	Nd	0.8	19.2	2.2	1.6
V	1.6	22.7	2,1	3.4	Sm	0.8	18.8	2.1	2.2
Cr	1.6	22.7	2,1	3.4	Gd	0.8	20.4	2.2	1.7
Fe	1.3	21.7	1.8	3.4	Tb	0.8	19.6	2.2	2.1
Co	1.2	21.8	1.7	3	Dy	0.8	19.8	2.3	2.4
Ni	1.2	21.7	1,8	3	Ho	0.9	20	2.3	2.4
Cu	1.3	22.7	2	3.5	Er	0.9	19.2	2.3	2
Zn	1	17.1	1.2	2	Hf	1	23.3	2.9	3.6
Ge	1.1	20	1.4	2	Ta	1.2	28.6	2.9	3.7
Sr	1.2	18.5	2	2.5	W	1.1	28.7	3.1	4.3
Y	1.3	19.9	2,3	2.4	Ir	1.1	25.5	2.7	4.2
Zr	1.5	22.7	2.6	3.7	Pt	0.8	23.7	2.1	4
Nb	1.6	27.9	3	4	Au	0.7	19.7	2	4
Mo	1.7	29.5	3.1	4.5	Pb	0.6	17.3	1.6	2
Rh	1.5	23.8	1.8	4.5	Bi	0.5	14.4	1.2	1.8
Pd	1.2	23.5	1.9	3.5	Th	1.0	23.3	2.9	2.4
Ag	1.1	22.8	2.1	4	U	1.1	23.5	2.3	3.4

5. Analysis of experimental results

There is reason to believe that the directional velocities of ions of lower charge decrease with increasing pressure more substantially due to collisions of the ions with the neutrals of the fill gas. At these collisions, owing to the nonresonance charge exchange of the ions of the cathode material with the gas neutrals, there occurs a decrease in ion charge. Thus, the singly charged ions are ions that have experienced more collisions with neutrals and consequently have a lower kinetic energy and a charge. Indirect evidence in support of this supposition is the experimental fact that the total ion current decreases as the pressure is increased due to the supply of gas into the discharge gap and that the effect of deceleration of ions is more pronounced if the fill gas has a greater specific mass.

The magnetohydrodynamic two-fluid model of the cathode flare plasma consisting of electrons and ions of medium charge number <Q>, which expands adiabatically after explosive formation of an "ecton" has been developed by G. A. Mesyats and co-workers

[15–17]. According to the results of their investigations [17], at a great distance from the cathode spot, the velocity of expansion of a cathode flare equal to the directional velocity of ions, v_i, can be determined as

$$v_{\mathrm{i}} = (2/(\gamma - 1)) \cdot \sqrt{\gamma(<Q>T_{\mathrm{e}} + T_{\mathrm{i}})/M_{\mathrm{i}}} \, , \tag{2}$$

where γ is the adiabatic exponent equal to 5/3. Putting, as in [17], $T_e \approx T_i$ and substituting into expression (2) the values of $<Q>$ and T_e from Table 1, we can estimate the directional velocities of ions. The resulting estimates are compared with experimental data in Fig. 6. The rather high degree of agreement between the calculations and measurements allows the conclusion that the observed ion velocities are achieved mainly due to the magnetohydrodynamic mechanism of acceleration, and expression (2) explains the experimental dependence of v_i on $<Q>$ and M_i as well as (since $T_e \propto V_{\mathrm{d}}$) on V_{d}. The magnetohydrodynamic model of the acceleration of ions in the cathode region of a vacuum arc [8], specially proposed to interpret the results of the present investigations, is based on the idea that the acceleration of the cathode material goes on after its transition to the state of a completely ionized collisional plasma. As in [15–17], it is assumed that the reason for the acceleration of ions is the expansion of the plasma into vacuum under conditions of constant inflow of energy due to Joule heating. By comparison of the model predictions with experimental data it has been shown that the acceleration of the cathode material occurs in the main when the material is in the state of ideal completely ionized plasma, through its expansion into vacuum, and the velocities of ions at a considerable distance from the cathode spot of the vacuum arc given by

$$v_{\mathrm{i}} = 3{,}5 \cdot \sqrt{\gamma <Q> T_{\mathrm{e}} \;\; M_{\mathrm{i}}} \, . \tag{3}$$

The predicted values of velocities are also given in Fig. 6. Good correlation with measurements and with the calculations using expression (2) is observed. An important feature of these investigations is the opportunity to calculate directional velocities for ions of various charge states. Based on the calculations it has been established that the velocity of ions of various charge states can differ from their common magnetohydrodynamic velocity only by a few percent, which is confirmed by the obtained experimental data.

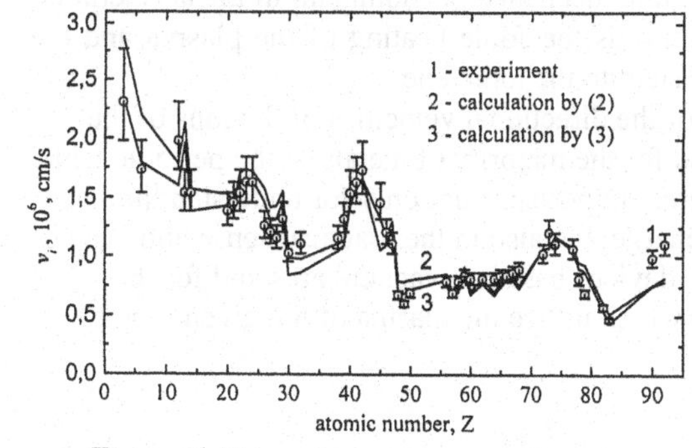

Figure 6. The predicted values of velocities.

In conclusion it should be noted that, according to the above magnetohydrodynamic models, the factors that determine the directional velocity of ions should be the mass of the ions, their mean charge, and the electron temperature, and this is in complete agreement with the obtained experimental results.

The experimental investigations of directional

velocities performed allow us to obtain an unambiguous relation between the density of the saturation ion current onto a flat Langmuir probe located perpendicularly to the plasma flow and the parameters of the plasma:

$$j_i = e<Q>nv_i = e<Q>n\,(20\,T_e<Q>/M_i)^{1/2}. \qquad (4)$$

The use of expression (4) is more justified than the frequently used (see, e.g., [13], page 344) estimation of the plasma parameters based on the well-known Bohm formula [18] derived on the assumption that the ions of a vacuum arc move with the sound velocity. At the same time, from the experimental results obtained it follows that the directional velocities of ions exceed substantially the ion sound velocities. It can easily be shown that the use of Bohm's formula for the definition of the plasma density by the ion current onto a flat probe gives a value underestimated more than three times.

6. Conclusions

The investigation results presented in this paper can be summarized as follows:

1. With the use of the emission method for measuring the directional velocities of ions of the vacuum arc plasma it has been shown that these velocities have values of the order of 10^6 cm/s, depending on the cathode material and practically identical for differently charged ions. The working gas supplied into the discharge gap reduces the directional velocities of ions, and this decreased is more pronounced for lower charges.

2. Based on the results of investigations carried out for a variety of cathode materials, it has been shown that the value of the steady-stated operating voltage of a vacuum arc discharge determines the directional velocities of ions. The established experimental facts in combination testify to the development of the processes of acceleration of ions in a small near-cathode region where an intense energy exchange occurs both between the electronic and ionic subsystems of the plasma and between variously charged ions.

3. Based on the experimental and predicted values of the directional velocities of ions, comparing the measured and predicted dependencies of the velocities the mass of ions for various materials and the mean charge of the ions and the electron temperature, one can state that the magnetohydrodynamic mechanism is dominant in the acceleration of ions, and the reason for the acceleration is the Joule heating of the plasma and the energy transfer from its electronic subsystem to the ionic one.

4. Comprehensive measurements of the directional velocities of the ions of the vacuum arc plasma have been performed for the majority of metals of the periodic table. The experimental data obtained are of great importance not only for understanding the physical nature of the given type of discharge, but also in the practical sense and can be used both in the development of various devices based on vacuum arcs and for the optimization of technological processes which utilize the plasma of the given type of discharge.

Acknowledgments

The author extends his sincerest appreciation to E. M. Oks, A. S. Bugaev, V. A. Gushenets, and A. G. Nikolaev (Laboratory of Plasma Sources of the Institute of High Current Electronics, RAS) as well as to I. Brown and A. Anders (Lawrence Livermore National Lab., Berkeley, USA) and to I. A. Krinberg (Irkutsk State University) for their fruitful and constructive discussions of the investigation results and help in the performance of experiments.

The work has been supported in part by research contracts with Lawrence Livermore National Laboratory, Berkeley, USA, under the IPP Trust-1 American-Russian scientific cooperation program and by the Russian Foundation for Fundamental Research grant No. 99-02-18163.

References:

1. J.M. Lafferty (ed.), *Vacuum arcs – theory and application.* - Wiley, New York, 1980.
2. G.A. Mesyats *Ectons* in a vacuum discharge: breakdown, the spark, and the arc.– Moscow, Nauka, 2000.
3. A.A. Plutto,V.N. Ryzhkov, and A.T. Kapin, *Zh. Eksp. Teor. Fiz.* **47,** 494 (1964).
4. E. Kobel, *Phys. Rev.* **36,** 1636 (1930).
5. W.D. Davis, H.C. Miller, *J. Appl. Phys.* **40,** 2212 (1969).
6. K. Tsuruta, K. Sekiya, and G. Watanabe, *IEEE Trans. Plasma Sci.* **25,** 603 (1997).
7. A.S. Bugaev, V.I. Gushenets, A.G. Nikolaev, E.M. Oks, and G.Yu. Yushkov, *Technical Physics.* **45,** 1135 (2000).
8. G.Yu. Yushkov, A.S. Bugaev, I.A. Krinberg, and E.M. Oks, *Doklady Physics.* **46,** 307 (2001).
9. I.G. Brown, J.E. Galvin, R.A. MacGill, and R.T. Wright, *Rev. Sci. Instrum.* **58,** 1589 (1987).
10. A.S. Bugaev, V.I. Gushenets, A.G. Nikolaev, E.M. Oks, and G.Yu. Yushkov, *Izv. Vyssh. Ucheb. Zaved., Fizika,* **2,** 21 (2000).
11. G. Yushkov, Proc. 19th Intern. Symp. on Discharges and Electrical Insulation in Vacuum. Xi'an, China, 2000, V. I, p. 260.
12. G.A. Mesyats *Ectons* – Ekaterinburg, Nauka, 1993.
13. I.G. Brown (Ed.) *The physics and technology of ion sources.* - Wiley, New York, 1989.
14. A. Anders, *Phys. Rev. E* **55,** 969 (1997).
15. E.A. Litvinov, In: *High-power nanosecond pulsed sources of accelerated electrons.* Ed. by G.A. Mesyats. - Novosibirsk, Nayka, 1974.
16. E.A. Litvinov, G.A. Mesyats, and D.I. Proskurovsky, *Sov. Phys. Usp.* **15,** 102 (1975).
17. G.A. Mesyats, and D.I. Proskurovsky, *Pulsed electrical discharge in vacuum.* – Springer-Verlag, Berlin, 1989.

GASEOUS PLASMA PRODUCTION USING ELECTRON EMITTER BASED ON ARC DISCHARGE

M.V.Shandrikov, A.V.Vizir, G.Yu.Yushkov, E.M. Oks
High Current Electronics Institute SD RAS
4 Akademichesky ave, Tomsk, Russia.

Abstract. Results of experimental study of low-pressure, high-current gaseous discharge with electron emitter based on arc discharge are presented. It is shown that the discharge of this type can be used for efficient generation of dense and uniform gaseous plasma. Special features of the electron emitter design allowed to reach long lifetime due to reduction the arc cathode erosion rate as well as to avoid contamination of the gas plasma by metal species. Plasma parameters measured by Langmuir, double and emissive probes are presented. High efficiency of energy utilization, wide range of operating pressure, reliability, and possibility to operate in chemically active environment favour the use of such a system in various surface modification processes.

1. Introduction

Low-pressure glow discharge is characterized by stability and uniformity of plasma parameters, as well as by high ion fraction of cathode current. That provides its applications in various ion sources [1] and plasma guns [2]. Despite of a number of positive properties, drawback of a glow discharge is relatively high voltage, in comparison, for example, with arc. Hence, intense sputtering of the cathode reduces the lifetime of devices and contaminates the gaseous plasma with particles of the cathode material. Another factor that limits expanding applications of the glow is relatively high operating pressure and narrow range of its variation.

It is commonly known that in a glow discharge the main process of current transmission on the cathode is caused by secondary ion-electron emission. The coefficient of secondary ion-electron emission γ usually does not exceed 0.1 for self-sustained low-voltage discharge. This implies, that the increase of electron fraction of cathode current, even rather small in comparison with the discharge current, can essentially change the discharge parameters and in particular its voltage. Such increase can be carried out "artificially" by external injection of electrons into the cathode region [3, 4]. If the conditions for acceleration of these electrons in cathode potential fall are created, they will be indiscernible from electrons emitted by the cathode as a result of secondary emission. Such injection of electrons is equivalent to the increase of γ.

E. Oks and I. Brown (eds.),
Emerging Applications of Vacuum-Arc-Produced Plasma, Ion and Electron Beams, 115–122.

Possibility of variation the parameter γ allows to expand essentially the pressure operating range of the glow discharge due to expanding limit of operation pressure toward its low values, or from other hand to reduce the discharge voltage under convention pressure range. Based on method of electron injection to support a hollow cathode glow, several ion sources and plasma gun have been made [5]. In these devices injected electrons were generated in additional "keeping" hollow cathode glow. That allowed to reach 7 A DC of discharge current as well as to drive the hollow cathode glow with pressure as low as $3 \cdot 10^{-5}$ Torr. Further current increase was limited by overheating the electrode of the glow keeping discharge and its intense sputtering; as well as by instabilities of the discharge parameters caused by time-to-time transitions of the keeping glow discharge to arc.

In this paper presented are the results of investigation of hollow cathode glow with electron injection where to generate electrons a constricted arc discharge is used.

2. Experiment

A schematic view of the electrode system is shown on Fig. 1. The plasma is generated by high-current, two-stage gaseous discharge. First stage (keeping discharge) is a filamentless electron emitter based on constricted arc. Second stage (main discharge) is a low-voltage low-pressure non-selfsustained glow discharge.

Cathodic spots are formed on the internal surface of the hollow cathode of the keeping discharge 1. For prevention of cathode spot formation on external surface of the cathode 1, it is covered outside with stainless steel shield that is insulated from the cathode. The electron flow moves to the grid-like keeping discharge anode 2 through the slot of the cathode 1 and slot-like constricting channel in the intermediate electrode 3. The use of slot shape of the constricting channel instead of an aperture allowed increasing the lifetime of the keeping discharge anode grid due to reduction of electron current density at the grid.

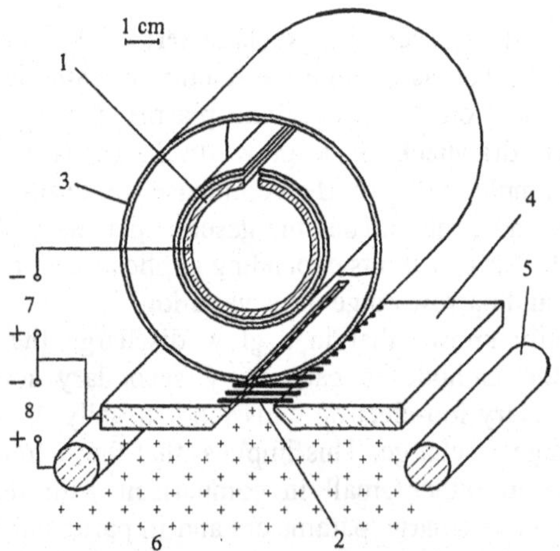

Figure 1. Schematic view of the plasma source. 1 – keeping discharge cathode, 2 – keeping discharge anode, 3 – intermediate electrode, 4 – main discharge cathode, 5 – main discharge anode, 6 – main discharge plasma, 7, 8 – power supplies of keeping and main discharges.

The anode of keeping discharge is electrically connected to the cathode of the main discharge. The greater part of electrons is extracted from the keeping discharge plasma through the electrode 2 by the cathode potential drop adjacent to the cathode 4. Injected electrons accelerated by the potential drop effectively ionize operating gas all over the vacuum chamber. Anode of the main discharge collects slowed injected electrons as well as main discharge plasma electrons.

The potential barrier for the electrons forms near the chamber walls, provided that all electrodes of the sy stem are insulated from the grounded chamber. Measurements show that the plasma potential sets to value about +20 V relatively to the chamber. At the same time, the main discharge voltage, and, correspondingly, the injected electron energy, is 100 V (eV). Despite of that, the potential barrier is enough to reflect some electrons capable of ionization, because injected electrons loose energy passing through the chamber, and, also, due to elastic collisions, the incidence angle of electron trajectory can differ from normal. The chamber walls are not subjected to intense sputtering that could contaminate the plasma, because of low ion energy (20 eV). Moreover, insulation of the electrode system from ground potential allows to avoid cathode spot formation on the chamber walls.

Cathodic spot on the internal surface of the cathode is initiated by ceramic surface breakdown (trigger unit is not shown). The configuration and arrangement of the cathode 1 and intermediate electrode 3 (Fig. 1) almost completely excludes contamination of the vacuum chamber volume by products of the keeping discharge cathode erosion. Hollow shape of the cathode, compared with an open cathode, provides longer cathode lifetime due to re-deposition of the cathode material on the opposite cathode wall.

The device operates in DC mode. All electrodes of the source are water-cooled. They are arranged within a vacuum case mounted on the chamber flange. The operating gas is fed into the keeping discharge cathode cavity. The pressure drop on the cathode and intermediate electrodes allows the discharge to operate with relatively low gas flow rate. Values of pressure noted in following sections refer to the pressure in the vacuum chamber. Operating gases were Ar, N_2, O_2.

The chamber volume is 0.9 m^3. There is a probe holder in the chamber that can be moved along the chamber axis perpendicular to the direction of electron injection. Plasma parameters were measured with Langmuir probe, double probe and emissive probe. Langmuir probe has an additional electrode serving to keep flat the emission surface of the plasma. Total surface area of the probe and additional electrode is 200 cm^2. Double probe is a pair of insulated equal plates made from stainless steel. The plates were placed on the probe holder. The voltage between plates was varied from 0 to 200 V. The power supply was insulated from ground. The current passing from one plate to another through the plasma was measured. The plates must be sufficiently large for emission surface of plasma to be flat. From other hand, the probe must be small for good spatial resolution. The surface area of a plate was chosen to be 28 cm^2. The results obtained with stainless steel and copper double probes were similar. At the same time, measurements made by aluminum double probe gave other results, and also these results were changing from time to time, obviously, because of presence of aluminum oxide film covering the surface of the probe. Emission probe was made from tungsten wire of 0.05 mm diameter and 5 mm length, mounted at the edge of ceramic tube of 3 mm diameter. The wire was heated by current of 1 A.

3. Results and discussion

Parameters of the constricted arc keeping discharge do not much differ from those reported elsewhere [6]. The discharge voltage slightly reduces with pressure (Fig. 2), and with current. If the pressure is higher than $6 \cdot 10^{-4}$ Torr, the lower limit of the constricted arc discharge current is determined by the current threshold of cathode spot operation, that depends on the cathode material [7]. With lower pressure, the discharge current becomes limited by the current transmission through the constricting channel. Stable initiation and enduring operation without extinguishing of constricted arc discharge required the open circuit voltage of the discharge power supply to be higher than 300 V, that is much higher than the discharge voltage. However, the power supply does not have to provide full current with this high voltage. Additional power supply providing 1 mA at 500 V, was connected in parallel to the main constricted arc power supply (10 A, 100 V).

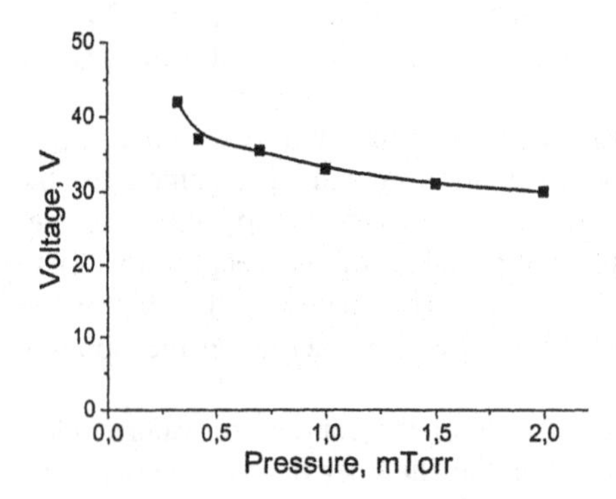

Figure 2. Pressure dependence of keeping discharge voltage. Discharge current is 7 A. Argon.

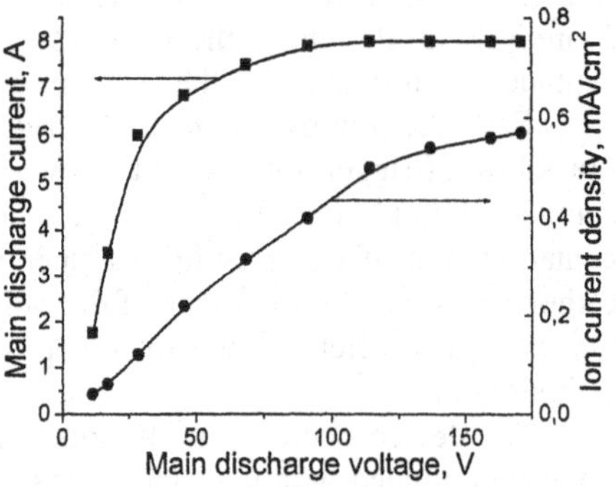

Figure 3. Dependence of the main discharge current on its voltage and corresponding variation of the ion current density. Probe bias voltage is –200 V. Keeping discharge current is 8 A. Pressure is 0.4 mTorr of argon.

Fig. 3 shows current-voltage characteristics of main discharge for constant keeping discharge current and corresponding effect of main discharge voltage on the ion current density measured by Langmuir probe. Between 20 and 60 V, rapid growth of the discharge current and density of its plasma occurs, caused by the increase of electron energy and ionization cross-section. Further voltage increase does not lead to current growth because electrons accelerated by the cathode potential drop of the main discharge start to escape to chamber walls. Nevertheless the ion current density slightly increases. Maximum efficiency of plasma generation is reached with main discharge voltage of 100-120 V.

Ion current extracted to the probe and, consequently plasma density are proportional to the discharge current (Fig. 4). The keeping discharge current lower than 2 A was reached by realizing hollow cathode glow discharge with the same electrodes as for constricted arc. The dependence does not have a feature that characterizes an electron beam plasma discharge, such as discontinuous behavior. Hence, the plasma is generated exceptionally due to individual particle interaction.

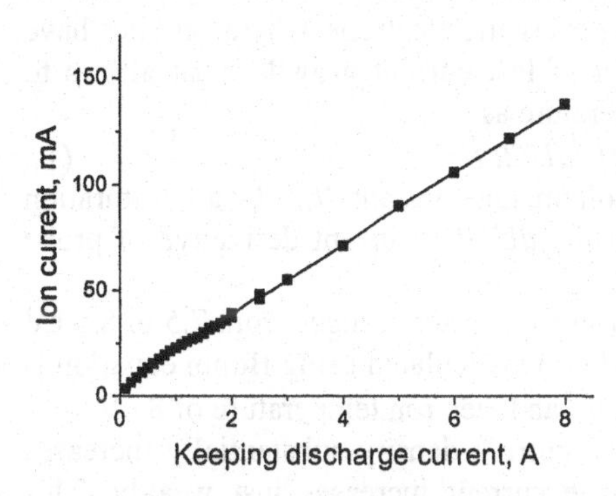

Figure 4. Ion current as a function of discharge current. Sample bias voltage is –200 V. Pressure is 0.6 mTorr of argon. Main discharge voltage is 115 V.

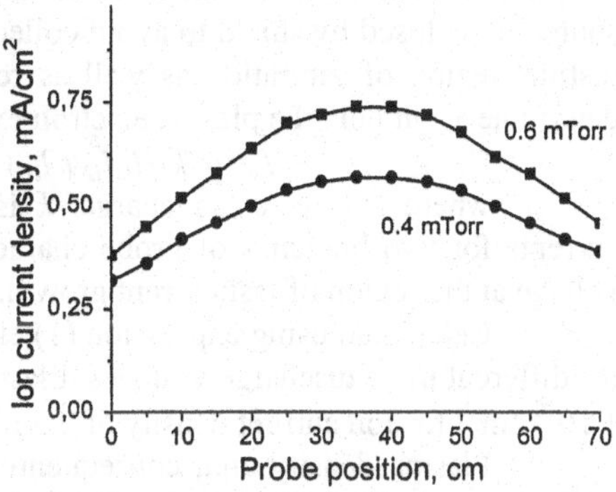

Figure 5. Ion current density distribution measured along the chamber axis. Source-to-axis distance is 55 cm. Argon. Keeping discharge current is 8 A. Main discharge voltage is 100 V.

Visually, plasma density is distributed sufficiently uniform over the chamber volume. That was confirmed by ion current density distribution measurements (Fig. 5) performed by moveable Langmuir probe. Because the cross-section of electron scattering on gas substantially exceeds that for inelastic collisions, injected electrons partially loose their original direction at the distance from the source about several tens of centimeters. The fact that electrons are scattered on the gas atoms much faster than they loose their energy to ionize atoms leads to a good uniformity of the plasma. Also, "shadow" effect was insignificant. Ion current density at the back side of the holder was just 16% less than at the holder side facing the source.

Despite that an arc is characterized by certain instability, temporal variation of ion current extracted from plasma was less than ±5%, obviously, due to arc constriction.

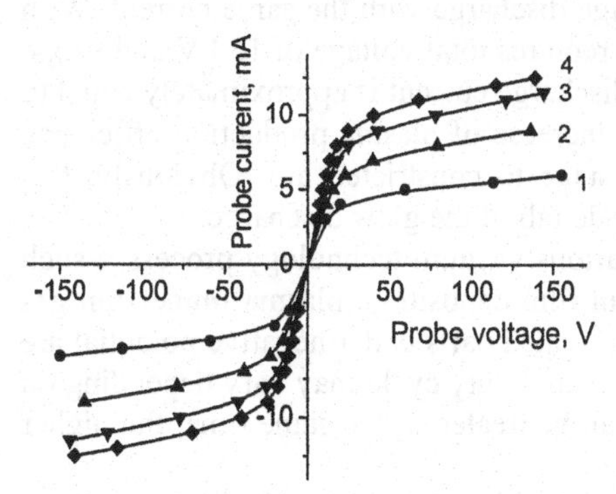

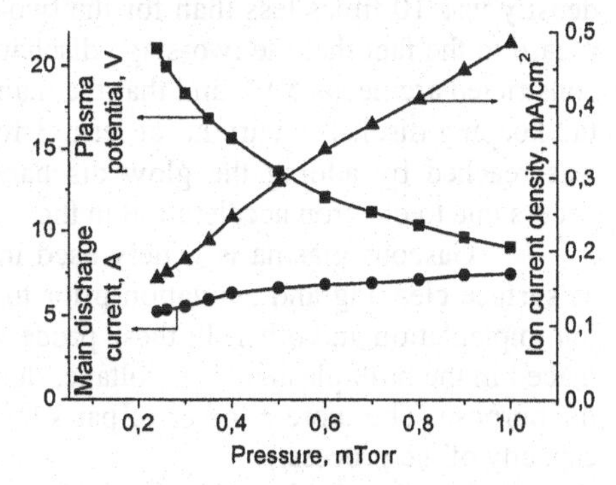

Figure 6. Double probe current-voltage characteristics. Keeping discharge current is 8A. Pressure is 0.5 mTorr of argon. 1 – main discharge voltage is 50 V, 2 – 80 V, 3 – 120 V, 4 – 150 V.

Figure 7. Pressure dependencies of main discharge current, density of ion current extracted from plasma, and plasma potential. Keeping discharge current is 7 A. Main discharge voltage is 115 V.

The double probe current-voltage characteristics (Fig. 6) are equal for different probe directions, in case that the collecting surface does not face the electron flow from the plasma source. To provide these conditions, the side of the probe facing the plasma source was closed by shield to avoid collecting energetic electrons. Characteristics have distinct region of saturation as well as region of fast current growth. That allows to determine accurately the plasma electron temperature as

$$T_e=(e/k)[I_{i1}I_{i2}/(I_{i1}+I_{i2})]\,/dU/dI\,/ \quad (1),$$

where e is electron charge, k is Boltzmann constant, I_{i1}, I_{i2} are saturation currents for two branches of probe characteristic, dU/dI – current derivative of probe voltage at the region of fast current growth.

Calculated using expression (1) electron temperature ranges from 7.5 to 8.5 eV for different main discharge voltages. Plasma density calculated using Bohm equation is $1 \cdot 10^{10}$ cm^{-3}, for ion current density of 0.5 mA/cm^2 and electron temperature of 8 eV.

Plasma density and, consequently, ion current density substantially increases with pressure (Fig. 7), though, main discharge current increases just weakly. This occurs due to reduction of ionization length of injected electrons that, in turn, causes the increase of electron energy fraction that is used for ionization. With low pressure, energetic electrons start to escape to chamber walls. The plasma becomes charged more positively. The increase of potential barrier near the chamber walls partially compensates electron loss.

A constricted arc itself is widely used for plasma generation in different devices, for example, ion sources [8]. As it was shown above, electron injection into glow discharge plasma substantially improves the glow discharge parameters. On the other hand, adding the "glow discharge stage" to a constricted arc essentially increases efficiency of plasma production. To compare efficiency of two-stage discharge and constricted arc discharge, the following test has been made. The grid-like anode 2 (Fig. 1) of the keeping discharge was removed, and the constricted arc power supply voltage was applied between cathode 1 and grounded anode 5. Thus, typical constricted arc electrode configuration widely used for bulk plasma production was realized. Despite the discharge plasma was filling the whole volume of the vacuum chamber, the plasma density was 10 times less than for the two-stage discharge with the same current. With regard to the fact that the two-stage discharge requires total voltage of 150 V and single constricted arc needs 50 V and that the main discharge current is approximately equal to the keeping discharge current, at least 3-fold increase of plasma production efficiency was reached by adding the glow discharge stage to constricted arc. Obviously, this occurs due to electron acceleration in the cathode fall of the glow discharge.

Gaseous plasma is widely used in various vacuum technology processes such as surface cleaning and activation prior to thin film deposition, plasma immersion gas ion implantation and other. In these processes, treated parts under negative potential are placed in the bulk plasma. The voltage, duration and duty cycle may vary depending on the purpose. The more number of parts that can be treated at the same time, the higher capacity of the process.

In general, plasma generation process may deteriorate as a result of putting a part with considerable surface area. However, extracting ions onto a large surface from the plasma of the glow discharge with external electron injection should not get worse the plasma parameters, because ionizing electrons are reflected by potential drop adjacent to the surface of the treated parts, and, moreover, electrons emitted by negatively biased parts contribute to plasma generation process.

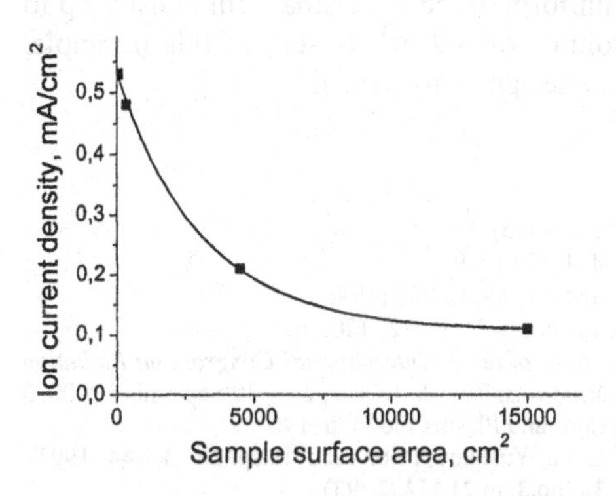

Figure 8. Ion current density as a function of the collector surface area. Keeping discharge current is 7 A. Main discharge voltage is 100 V, pressure is 1 mTorr of argon.

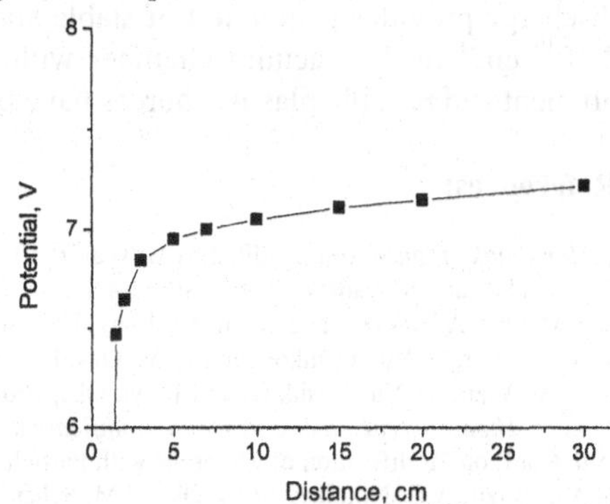

Figure 9. Potential as a function of distance from negatively biased collector. Collector surface area is 5000 cm^2. Bias voltage is 200 V. Discharge current is 3 A.

Fig. 8 shows the dependence of surface-average density of ion current extracted onto a collector on the collector surface area. The density of ion current for large collector is substantially less than that for a small collector. One of the reasons is the fact that a significant part of large collector is located away from the region of maximum plasma density. Nevertheless, according to given above spatial ion current density distribution, at the edge of the collector with 4500 cm^2 the ion current density was just 30 % less than maximum, while the surface-averaged current density reduces in a factor of 2.5 compared with the collector of 60 cm^2. One more reason can be a presence of weak electric field in the plasma that accelerates ions to the collector. If the collector is small, this electric field focuses the ion flow to the collector. Measurements performed by emissive probe show that such a field actually exists near the collector under negative potential (Fig. 9). The field extends to the distance that is much greater than the length of positive space charge layer. Since the plasma is charged positively, such a field must be present near the chamber walls. That was also confirmed by measurements. Potential difference within the plasma is about several tenth of volt. To answer the question if this field is enough to give ions direct drift velocity, additional experiments are required.

Which ion current density must be taken to correctly determine the plasma density using Bohm equation, for small collector or for large, is a question. Note that reported above plasma density ($1 \cdot 10^{10}$ cm^{-3}) was defined using commonly used technique with small (200 cm^2) Langmuir probe.

Recently, a new version of plasma source based in the same principle as described above has been fabricated and tested. The new plasma source operates with keeping discharge current up to 20 A. The ion current density reaches 1.5 mA/cm^2. Corresponding plasma density is $3 \cdot 10^{10}$ cm^{-3}.

Conclusion

The use of constricted arc-based emitter for electron injection into the glow discharge provides generation of stable and uniform gaseous plasma with density up to $3{\cdot}10^{10}$ cm^{-3} in the vacuum chamber with volume of 0.9 m^3. Based on this principle, efficient and reliable plasma sources have been designed and tested.

References:

1. *Handbook of ion sources.* Edited by B. Wolf (CRC Press, 1995)
2. A. Anders and S. Anders, Plasma Sources Sci. Technol. **4**, 571 (1995)
3. E.M. Oks, A.V. Vizir, and G.Yu. Yushkov, Rev. Sci. Instrum. **69(2)**, 853 (1998)
4. A. V. Vizir, G. Yu. Yushkov, and E. M. Oks, Rev. Sci. Instrum. **71(2)**, 728 (2000)
5. A. V. Vizir, G. Yu. Yushkov, and E. M. Oks, *Proceedings of the 1st International Congress on Radiation Physics, High Current Electronics, and Modification of Materials,* Tomsk, Russia, 24 – 29 September 2000, **3** (Vth Conf. on Modification of Materials with Particle Beams and Plasma Flow), p. 190.
6. A.V. Vizir, A.G. Nikolaev, E.M. Oks, P.M. Schanin, G.Yu. Yushkov, Prib. Tech. Eksp. No. **3**, 144 (1993). Translation: Instruments and Experimental Techniques, **36** (no.**3**, pt.**2**):434 (1993).
7. G.A. Mesyats *Cathode Phenomena in a Vacuum Discharge*. (Moscow, Nauka, 2000).
8. S.P. Bugaev, A.G. Nikolaev, E.M. Oks, P.M. Schanin and G.Yu. Yushkov. Rev. Sci. Instrum.1994. V.65, 10. P.3119-3125.

VACUUM ARC ION SOURCES: CHARGE STATE ENHANCEMENT AND ARC VOLTAGE

M. GALONSKA, F. HEYMACH, R. HOLLINGER AND P. SPÄDTKE
Gesellschaft für Schwerionenforschung mbH,
Planckstraße 1, D-64291 Darmstadt, Germany

Abstract. The Metal Vapor Vacuum ion source (MEVVA) has been developed for the production of ion beams for a wide range of metal ions. The ion charge states for most heavy elements created by the high current MEVVA ion source, which typically generates ions with a mean charge state up to three, have to be elevated to higher charge states for accelerator reasons at the GSI accelerator facility. Several methods are known which elevate charge states. Especially for the generation of U^{4+} ions a strong magnetic field and a high arc current has been used at GSI. Both methods increase the arc voltage which in turn influences the electron energy distribution and thus the mean ion charge state. Arc voltages have been measured for different settings of arc current, magnetic flux density and discharge geometry indicating the dependencies between these quantities and the arc voltage.

1. Introduction

The MEVVA ion source serves as a high current ion source for most conducting elements. The MEVVA ion source produces typically a charge state distribution with a mean charge state of about three [1,2]. However, at the GSI facility only ions with a mass to charge ratio up to 65 in atomic units can be accelerated. Taking an uranium beam for example, the suitable ion for accelerator reasons is U^{4+}. Since the fraction of U^{4+} produced by a MEVVA ion source is typically below 10 % without and about 30 % with strong magnetic fields efforts have been made to increase the fraction of U^{4+} ions. Low energy beam transport becomes easier in this case and beam losses are reduced [3]. Therefore the charge state within the plasma of the MEVVA ion source are enhanced.

Charge state enhancement has been performed by strong magnetic fields [4-7], by a high arc current [8], combining both methods [9,10] and by the "current spike" method [11]. Attempts have also been made by injection of an electron beam through a hollow cathode into a MEVVA discharge, the so called E-MEVVA [12,13]. At GSI an experiment has revealed that even an azimuthal magnetic field caused by a current through a wire which is guided through a hollow anode leads to a charge state shift [14]. By using a strong axial magnetic field of 180 mT and an arc current of 800 A a fraction of 67 % (current fraction) of U^{4+} ions is found within an uranium beam at GSI [15,16].

This shift of charge states is correlated with an increase of the arc voltage (sometimes addressed as burning voltage). The arc voltage influences the electron en-

E. Oks and I. Brown (eds.),
Emerging Applications of Vacuum-Arc-Produced Plasma, Ion and Electron Beams, 123–130.

ergy distribution thus causing charge state alteration. First assumptions may be that an increasing arc voltage causes an increase of the mean charge state. For this reason the arc voltage has been altered by different means (various arc currents, different magnetic flux densities and field arrangements, discharge geometries, anode materials) taking titanium as cathode material. With these investigations the relation between arc voltage and charge state enhancement can be discussed.

2. Experimental Setup

Measurements of the arc voltage were performed using a vacuum arc ion source using different discharge geometries which differ in distance between cathode and anode, anode aperture and anode material.

The discharge arrangement (see Fig. 1) has a tubular anode opposite to the cathode (Ti, 5.7 mm in diameter). The anode is partly shielded by the cathode flange (opening is up to 30 mm). The magnetic field configuration for charge state enhancement is created by two coils (coil 1and 2 in the drawing). These are connected in series and operated in pulsed mode. The first coil is placed at cathode position and the second one near the front face of the anode.

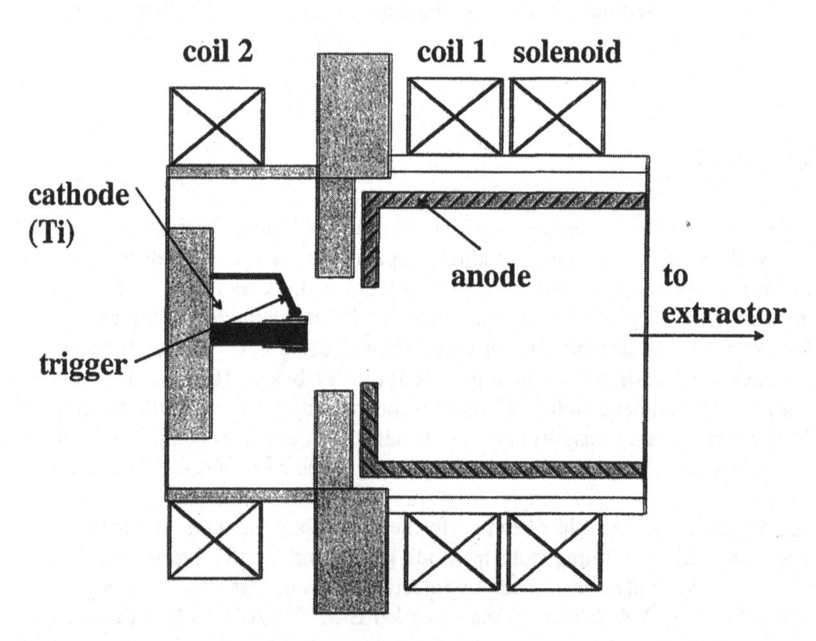

Fig. 1: Schematic drawing of the ion source (not in scale).

Both coils are entirely located outside the plasma chamber for vacuum reasons. Its magnetic fields are directed towards the extraction system of the source. A solenoid, which is placed in the back part of the tubular anode, is additionally installed. The tubular anode can easily be replaced by an anode of different aperture (10 to 25 mm in diameter), length (hence varying the distance between cathode and anode from 9 to 20 mm) or material for the investigation of the arc voltage as function of geometric and material changes.

3. Experimental Results

Fig. 2 displays a typical pulse shape of the arc voltage with 0.8 ms duration and a repetition rate of 1 pps respectively. The applied ("open circuit") voltage between cathode and anode is 50 V (left). During the build-up time of the plasma this voltage drops to the actual value of the arc voltage of 20 V (right). The superimposed noise of approx 15 % is directly transferred into ion beam fluctuations if no corrective measures are taken. This could be for example the use of meshes in the plasma channel [15].

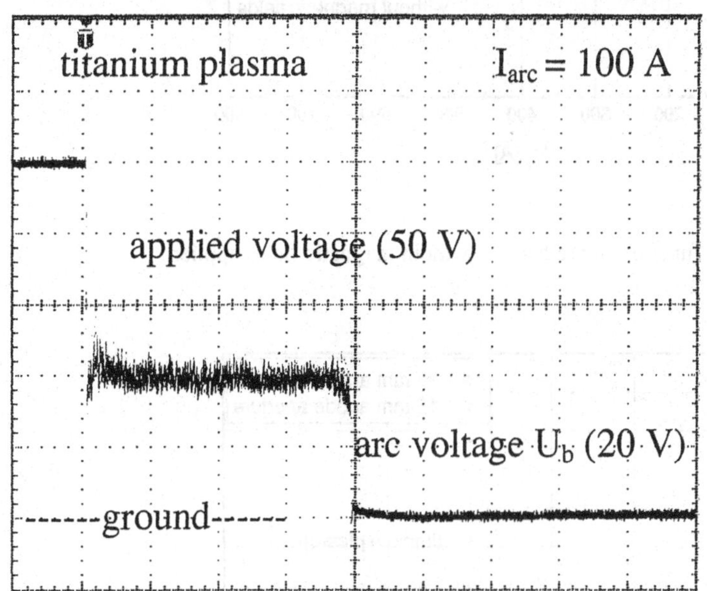

Fig. 2: Typical pulse shape of the arc voltage. Vertical: 10 V/div., horizontal: 200 μs/div.

3.1. ARC CURRENT AND ARC VOLTAGE

Fig. 3 shows the arc voltage U_b as function of the arc current I_{arc} without any magnetic fields applied. A copper anode is used, placed 15 mm apart from the titanium cathode. The influence of different anode apertures has been measured in addition for 15 and 25 mm aperture diameter.

In both cases the arc voltage increases with increasing arc current, but deviates from a linear curve. This is an evidence for a decreasing plasma resistivity with arc current. Regarding plasma resistivity Z_{plasma} versus arc power P_{arc}, there is in fact a decrease of resistivity observable (Fig. 4).

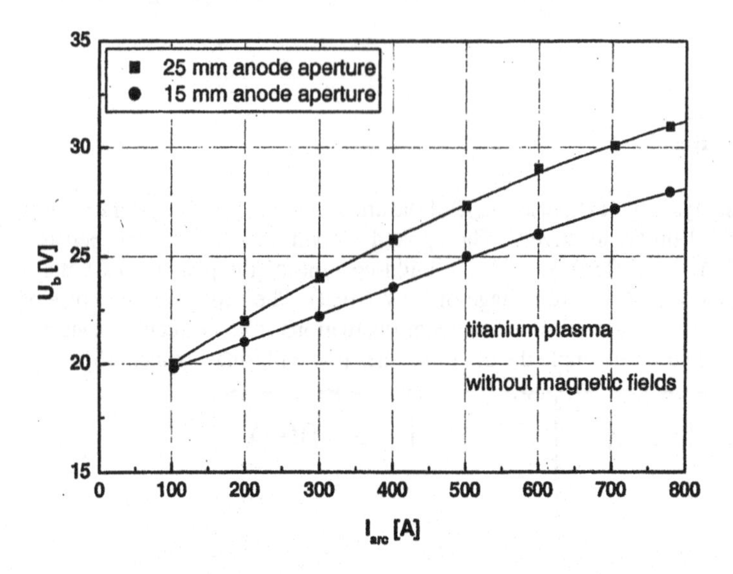

Fig. 3: Arc voltage as function of arc current without magnetic fields applied.

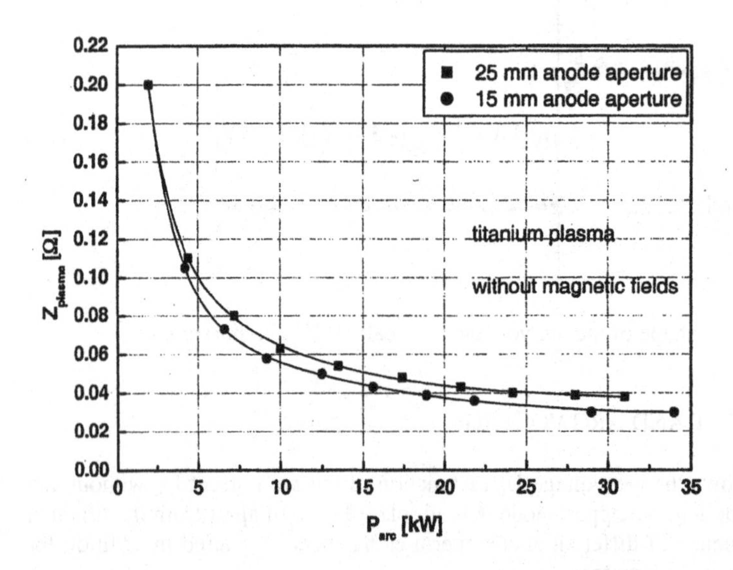

Fig. 4: Plasma resistivity as function of arc power without magnetic fields applied.

This could be due to an increase of electron density and of electron temperature taken into account the resistivity Z which is depending on the electron temperature by $T_e^{-3/2}$. On the other hand the arc voltage increases from 20 V at an arc current of 100 A to 31 V at a current of 780 A (anode aperture 25 mm). The spectrum is, however, only weakly sensitive to the increase of arc voltage in the absence of magnetic fields (see Fig. 7). While there is an increase of electron temperature, the mean electron energy which is responsible for a shift of charge states is only slightly enhanced by an increasing arc voltage without magnetic field.

There are indications that the anode geometry has an influence on the arc voltage and plasma resistivity which can be observed in Fig. 3 and 4. Because of a larger distance between center of cathode and anode electrode the resistivity is larger for an anode with larger aperture. Consequently the arc voltage rises.

3.2. MAGNETIC FIELD AND ARC VOLTAGE

Using a copper anode (aperture is 15 mm in diameter) the influence of a strong magnetic field B created by coils 1 and 2 on the arc voltage is indicated in Fig. 5.

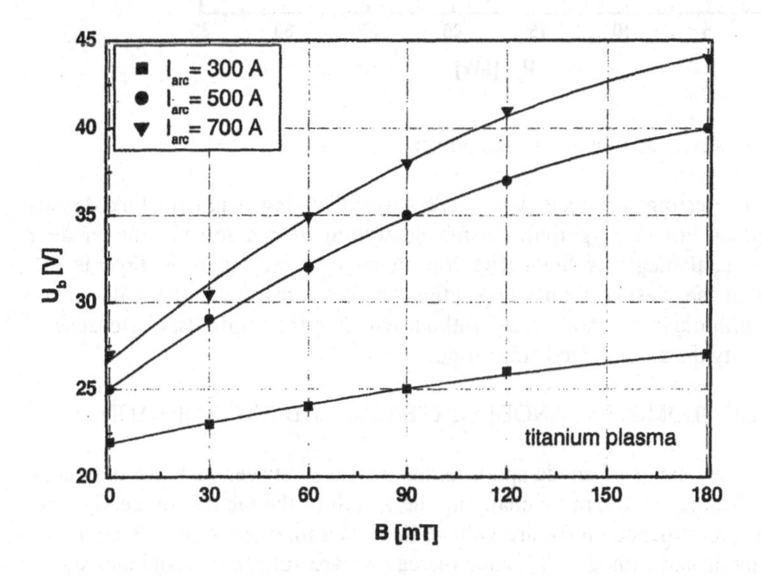

Fig. 5: Arc voltage as function of magnetic flux density (created by coils 1 and 2).

There is an increase of arc voltage with rising magnetic flux density (and of course with rising arc current as before). The effect of arc voltage increase is more pronounced for higher arc currents. This indicates that combination of magnetic field and arc current will have the strongest influence on plasma parameters. As a matter of fact the charge state distribution (expressed as mean charge state in Fig. 7) now is sensitive to an increasing arc voltage caused by a rising arc current. But this holds true only in the presence of magnetic fields. That means that in this case the rising arc voltage is transferred into an increasing mean electron energy as anticipated before.

In terms of plasma resistivity versus arc power (Fig. 6) the resistivity falls with rising arc power (see above).

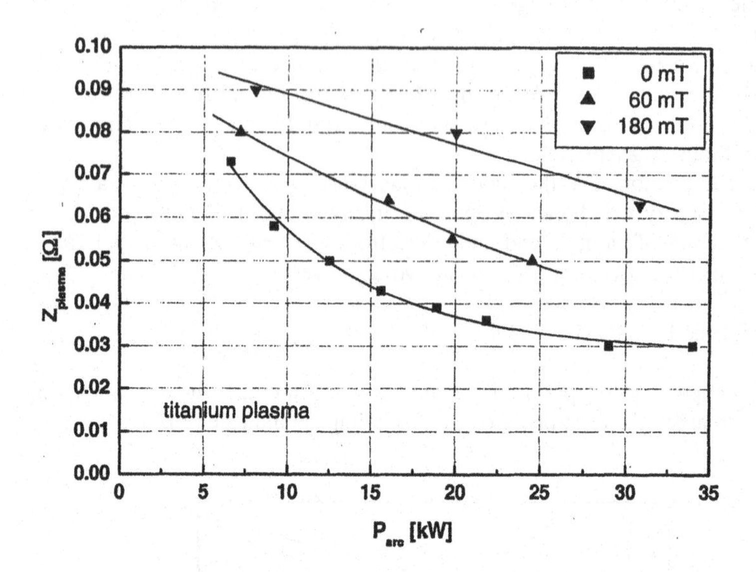

Fig. 6: Plasma resistivity as function of arc power.

More interesting, the resistivity increases with rising magnetic flux density (following a vertical line in the graph). Plasma electron movement towards the anode is impeded by the axial magnetic field. The confinement by the magnetic field is that strong that part of the plasma plume (depending on flux density) can enter the anode aperture, moving through the anode body without reaching the anode itself. Because of the larger resistivity the arc voltage is increased.

3.3. DISCHARGE GEOMETRY, ANODE MATERIAL AND ARC VOLTAGE

Besides the change of anode aperture, the distance between cathode and anode has been varied from 9 to 20 mm by changing the length of the tubular anode. The anode aperture has an influence on the arc voltage due to an increase of the effective distance between anode and cathode. The same increase of arc voltage with rising distance can therefore be anticipated. Curiously, no variation of arc voltage with distance has been observed neither with nor without magnetic fields. The distance only influences the ignition behaviour of the arc, the pulse-to-pulse stability and the noise which is superimposed to the arc voltage pulse. Making the distance too far (20 mm), arc ignition is getting instable. Making the distance closer (9 mm), arc voltage fluctuations occure to a huge extend in the presence of the magnetic field. Further surveys will be performed with a distance of 15 mm between cathode and anode.

Anticipating that the plasma resistivity falls with the number of charged particles in the spacing between cathode and anode (hence decreasing the arc voltage), the number of secondary particles created at the anode should influence the arc voltage.

This influence should be strong in the régime where anode spots occure, i.e. when using high arc currents (≥ 700 A). Different anode materials with different emission coefficients (for example secondary electron emission coefficient) should therefore change the arc voltage just as the arc voltage depends on the cathode material [17]. This comparison has been performed using anodes made of copper, aluminium and stainless steel. All of these have an aperture of 15 mm. The anode is placed 15 mm apart from the titanium cathode. The comparison displays that there is no difference in arc voltage between steel and aluminium independend of magnetic flux density and arc current. However, measured arc voltages, using a copper anode, are 5 V below the ones for Al and steel with arc currents of 500 A and 700 A. Since the secondary electron emission coefficient for Cu is above the ones for Al [18] these rough findings seem to be consistent with the anticipation above.

3.4. ARC VOLTAGE AND CHARGE STATES

In order to demonstrate the influence of the arc voltage on the spectrum increased by strong magnetic fields and high arc currents Fig. 7 shows the mean charge state (electrical) of a titanium plasma.

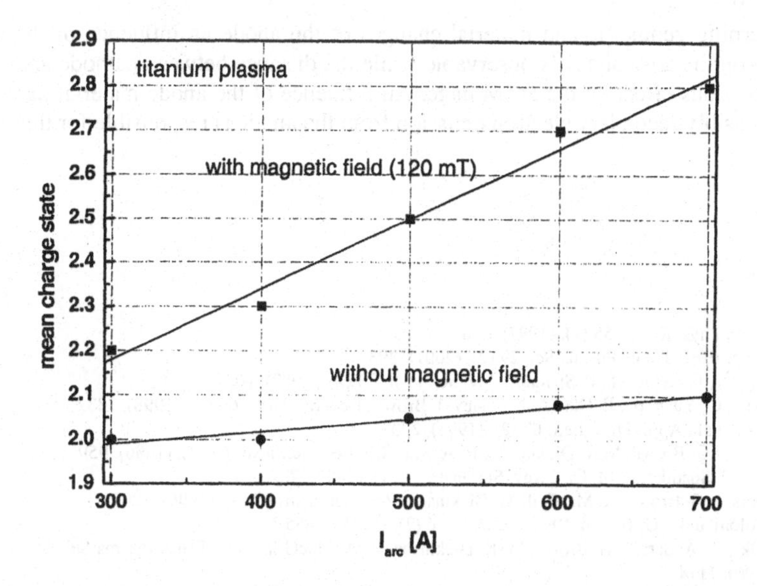

Fig. 7: Mean charge state as function of arc current with and without magnetic field (created by coils 1 and 2).

The mean charge state is displayed as function of arc current with and without magnetic field. For this survey a copper anode with an aperture of 25 mm in diameter was used. The distance between cathode and anode was kept at 15 mm. As stated above an increasing arc voltage in the absence of magnetic fields only slightly influences the spectrum. But evidently it does with increasing flux density. A high fraction of Ti^{3+} ions (about 60 % electrical, generated with a current of 700 A and a fluxc density of 180 mT) corresponds to an arc voltage of 41 V, which is suitable for the production of this ion species. At the same time a fraction of 10 % Ti^{4+} ions is observable.

4. Summary

Charge state enhancement by strong magnetic fields and high arc currents has been investigated in terms of arc voltage. Evidently both methods increase the arc voltage. The arc voltage in turn influences plasma parameters such as electron energy distribution, hence can lead to a shift of charge states. Therefore, the arc voltage is an important quantitiy when talking about charge state enhancement. On the other hand the correlation between arc voltage and charge state enhancement is ambiguous. As indicated, the relation between charge state distribution and arc voltage differs substantially for the cases with and without magnetic fields. This has to be regarded as a starting-point for further surveys on this topic. Measurements of the ion and electron energy distribution will be performed in the near future taking an electrostatic 127° cylinder spectrometer [19].

Concerning geometric and material changes of the anode an influence of the anode aperture on the arc voltage is observable while the distance between cathode and anode has no obvious effect on the arc voltage. An influence of the anode material has been shown. Possibly secondary electron emission from the anode is responsible for this behaviour.

5. References

1. A. Anders, Phys. Rev. E 55 (1), (1997), 696
2. A. Anders, IEEE Trans. Plasma Sci. 29 (2), (2001), 393
3. R. Hollinger, F. Heymach, P. Spädtke, Rev. Sci. Instr. 73 (2), (2002), 1024
4. A. Anders, G. Yushkov, E. Oks, A. Nikolaev, I. Brown, Rev. Sci. Instr. 69 (3), (1998), 1332
5. E.M. Oks et al., Appl. Phys. Lett. **67** (2), (1995), 200
6. E. M. Oks, I.G. Brown, M.R. Dickinson,, R. A. MacGill, Rev. Sci. Instr. 67 (3), (1996), 959
7. H. Reich, P. Spädtke, E. M. Oks, Rev. Sci. Instr. 71 (2), (2000), 707
8. A. Anders, I. G. Brown, R. MacGill, M. Dickinson, Rev. Sci. Instr. 67 (3), (1996), 1202
9. F. J. Paoloni and I. G. Brown, Rev. Sci. Instr. 66 (7), (1995), 3855
10. E. M. Oks, A. Anders, I. G. Brown, M. R. Dickinson, R. A. MacGill, IEEE Trans. Plasma Sci. 24 (3), (1996), 1174
11. A.S. Bugaev, E. M. Oks, G. Y. Yushkov, A. Anders, I. G. Brown, Rev. Sci. Instr. 71 (2), (2000), 701
12. A. Bugaev, V. Gushenets, G. Yushkov, Appl. Phys. Lett. 79 (7), (2001), 919
13. V. A. Batalin et al., Rev. Sci. Instr. 73 (2), (2002), 702
14. M. Galonska, F. Heymach, R. Hollinger, R. Lang, P. Spädtke, unpublished
15. R. Hollinger, M. Galonska, F. Heymach, P. Spädtke, (2002), MEVVA ion source for uranium high current operation at the GSI accelerator facility, to be published in the Proc. of the XXth ISDEIV, Tours, France
16. P. Spädtke et al., GSI-Scientific Report 2001, (2002), 191
17. A. Anders, B. Yotsombat, R. Binder, J. Appl. Phys. 89 (12), (2001), 7764
18. Stöcker, (1994), Taschenbuch der Physik, Verlag Harri Deutsch, Frankfurt am Main
19. R. Hollinger, K. Volk, H. Klein, Nucl. Instr. Meth. A 481, (2002), 86

LINEAR VACUUM ARC EVAPORATORS FOR DEPOSITION OF FUNCTIONAL MULTI-PURPOSE COATINGS

A.V. Demchyshyn[1], Yu.A. Kurapov[1], V.A. Michenko[1], Ye.G. Kostin[2], A.A. Goncharov[3] and Ye.G. Ternovoi[1]

[1] *E.O. Paton Electric Welding Institute of the National Academy of Sciences in Ukraine, 11 Bozhenko Str., 03680 Kiev, Ukraine*
e-mail: demch@iptelecom.net.ua

[2] *Institute for Nuclear Research of the National Academy of Sciences in Ukraine, 47 Nauka Av., 01028 Kiev, Ukraine*

[3] *Institute of Physics of the National Academy of Sciences in Ukraine, 46 Nauka Av., 01028 Kiev, Ukraine*

Abstract. Long, cylindrical cathodic arc evaporators (500 and 1550 mm long) with an external side-wall working surface are considered. The evaporators run in pulsed or continuous mode and allow deposition of functional coatings of pure metals, multi-component alloys, and refractory compounds on long and large items in order to increase their wear-, corrosion- or erosion-resistance. The surface microstructure and microhardness of corrosion-resistant FeCrAl coatings on high-temperature strengthened steel substrates, heat-resistant NiCrAlY coatings on high-temperature strengthened ЖС6У nickel alloy, copper coatings for brazing on 65Г steel, and erosion-resistant TiN coatings on B-95 aluminum alloy and BT6 titanium alloy are described and discussed. It is shown that vacuum-arc deposition of functional coatings using long evaporators provides the possibility of increasing significantly the working characteristics of critical machinery parts in modern engineering industry.

1. Introduction

In recent decades the widest developments of vacuum PVD technologies have been made with the use of ion-plasma techniques. These methods differ from other vacuum methods because of the presence an ion component in the vapor flow. This allows deposition of dense coatings with smaller coating thickness by means of ion energy control, synthesis of refractory compounds in reactive gases, and high adhesion of the deposited coatings to the substrates [1]. These technologies include vacuum arc evaporation of electrically conducting materials, characterized by a high degree of vapor ionization, adequate deposition rates of protective layers, good reproducibility of process parameters, inertialessness of the process, and high economic efficiency of the equipment.

E. Oks and I. Brown (eds.),
Emerging Applications of Vacuum-Arc-Produced Plasma, Ion and Electron Beams, 131–149.

At present, cathodic arc plasma sources utilizing the end face of a cylindrical cathode as the plasma generating surface ("butt-type") are mostly applied in practice. Such cathodic arc plasma sources are widely used in ion plasma coaters such as "Bulat", "Pusk", ВУ-2МБС, ННВ-6 etc. [2,3]. For the case of these "butt-type" systems for plasma generation, an outer coaxial electromagnetic system confines cathode spots and causes them to rotate along circular orbits on the cathode end surface. These sources generate a non-uniform, size-limited plasma flow that substantially restricts their applications.

However, the ever increasing requirements of the machine building industry put forward tasks for coating of long-sized and large-dimensional work-pieces that in turn require the production of cross-sectionally uniform plasma flow, and this cannot be realized by means of increase in the number of separate cathodic arc plasma point sources. Because of this, one of the urgent problems connected with the development of low pressure cathodic arc plasma sources is the necessity for the development of long linear electric arc sources with steered cathode spot motion on the side-wall working surface of a cylindrical cathode, and increased reliability of arc retention at the required erosion zone of the cathode.

Here we describe the design of a long cylindrical-type cathodic arc plasma source with a continuous mode of operation and with few of the disadvantages of the more conventional vacuum arc deposition systems.

2. Background

In recent years linear plasma sources with cathode spot magnetic stabilization have been developed [5]. In the case of long cylinder type pulsed arc evaporators the motion of the cathode spots is controlled by the magnetic field of the current channel within the plasma and the current that is conducted through the cathode [4,5]. In this case the magnetic field lines formed by these currents are tangential to the cathode surface. The cathode spots are moved from the trigger electrode towards the current lead-in wire, having both longitudinal (along the cathode) and azimuthal components. Thus the cathode spots undergo a rotary motion around the cylindrical cathode and move toward the current lead-in. When the cathode spots reach the arc-extinction screen they move into the gap between the cathode and a neutral screen that results in discharge extinction. The life-time of cathode spots on the cathode working surface is determined by the time for the spots to move to the arc extinction screen and depends on the discharge current value.

A design has been proposed [6] in which a long plasma source, working in pulsed mode, forms a directed tape type flow due to the presence of an outer magnetic system. A superimposed magnetic field confines the cathode spots from azimuthal movement due to its orthogonal field component, while the tangential component provides their longitudinal movement with a straight trajectory. The uniform distribution of ion current density along the cathode axis and the ordered mode of cathode spot motion along the cathode working surface provides the possibility of depositing coatings with an even distribution along the height (length) of a long workpiece.

A cylinder-type plasma source has been developed [7] with a radially spread plasma flow in order to improve the plasma source construction and to allow the possibility of increasing the number of items treated simultaneously. See Fig. 1

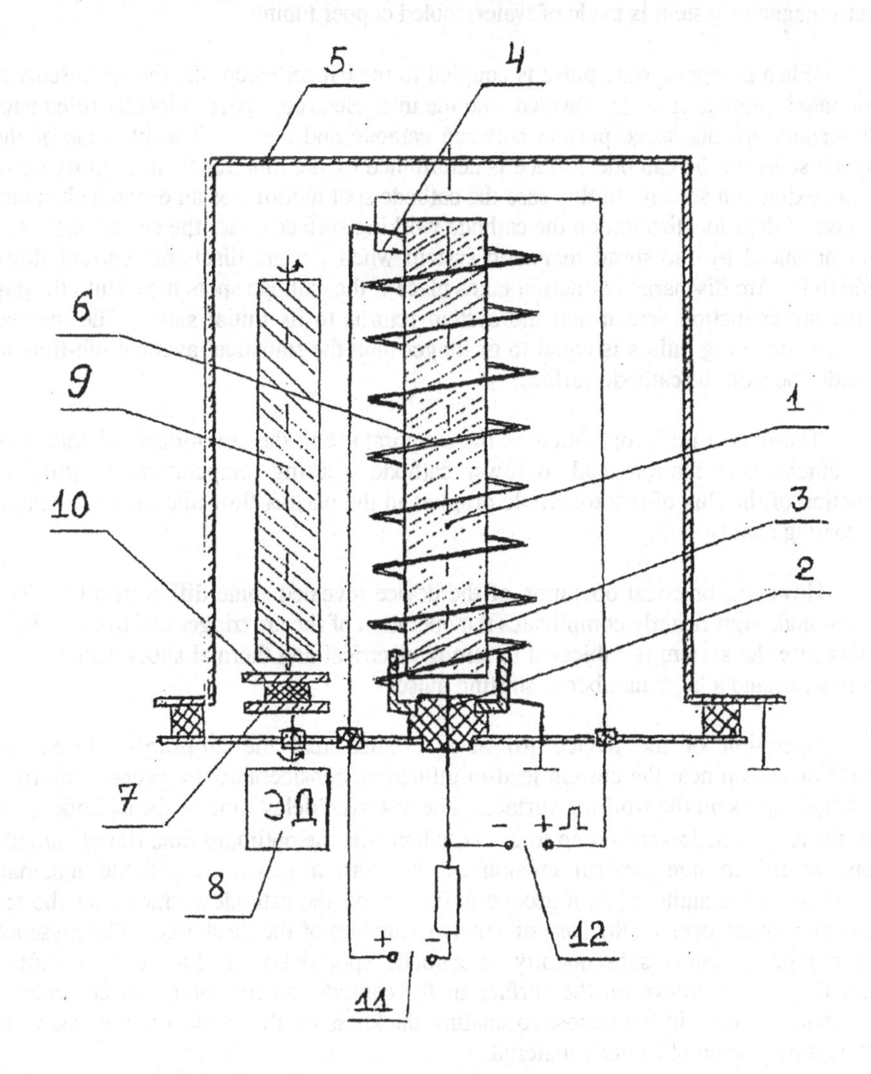

Figure 1. **Vacuum-arc cylinder-type evaporator [7]:**
1- cathode; 2 - current lead-in; 3 - arc-extinction screen; 4 - trigger electrode; 5 - anode; 6 - spiral magnetic system; 7,8 - rotation system with electric motor; 9 - workpiece to be coated; 10 - insulator.

The principle of action of this plasma source is based on control of the cathode spot motion by means of an outer magnetic field produced by a long solenoid system. This system is made in a helical shape whose spiral turns embrace the cathode working surface, and is mounted coaxially with the cathode (Fig. 1). In this case the negative side of the arc power supply is connected through a ballast resistor to the current lead-in of the long cathode. The positive pole is connected to the lead-in of the solenoid system on the trigger electrode side. It is at anode potential and grounded. The electromagnetic system is made of water-cooled copper tubing.

When an appropriate pulse is coupled to the trigger electrode, the arc discharge is initiated, cathode spots are formed, and the inter-electrode space is locally filled with the vacuum arc discharge plasma between cathode and anode. The life-time of the cathode spots on the cathode surface is determined by the time for them to move up to the arc extinction screen. In this case the cathode spot motion has an ordered character because of their localization on the cathode working surface under the curved magnetic field produced by the spiral magnetic system when the arc discharge current flows through it. Arc discharge extinction takes place if the cathode spots move into the gap of the arc-extinction screen and the system returns to its initial state. The interval between triggering pulses is equal to or longer than the statistical average life-time of cathode spots on the cathode surface.

The pulsed mode operation of the evaporator and the controllable character of the cathode spot motion lead to lower cathode working temperature, resulting in reduction of the flux of macroparticle droplets in the plasma flow and improvement in the coating quality.

However, practical operation of the device revealed some difficulties [5]. The pulsed mode significantly complicates the operation of the arc trigger electrode system. In this case the system is subjected to strong electrical and thermal shock loading and has to withstand a large number of starting pulses.

Operation of the device [6] also revealed that the arbitrarily placed arc extinction screen near the current lead-in influences considerable the over-all life-time of cathode spots on the working surface. The observed delay time of the cathode spots near the arc-extinction screen up to 2 - 4τ, where τ is the optimum time run of cathode spots, results in non-uniform erosion of the plasma generated cathode material. According to the authors [5], a groove is formed on the cathode surface near the arc extinction screen even in the case of timely extinction of the discharge. The presence of the groove results subsequently in cathode spot delay at this zone, a further deepening of the groove on the surface of the cathode material and, consequently, a substantial increase in the deposited coating thickness on the work-piece at this zone and poor utilization of cathode material.

In this report we describe a long cylindrical-type cathodic arc plasma source with a continuous mode of operation and with none of the disadvantages described above.

3. The New Long Linear Cathodic Arc Deposition System

The principle of action is also based on cathode spot motion steered by means of the magnetic field produced by a long solenoid; see Fig. 2. The system is made of copper tubing of helical shape. However in our case the tubular water-cooled cathode has arc-restriction screens instead of arc-extinction screens on both ends of the vacuum arc evaporator. Re-commutation of the applied electric load takes place when the cathode

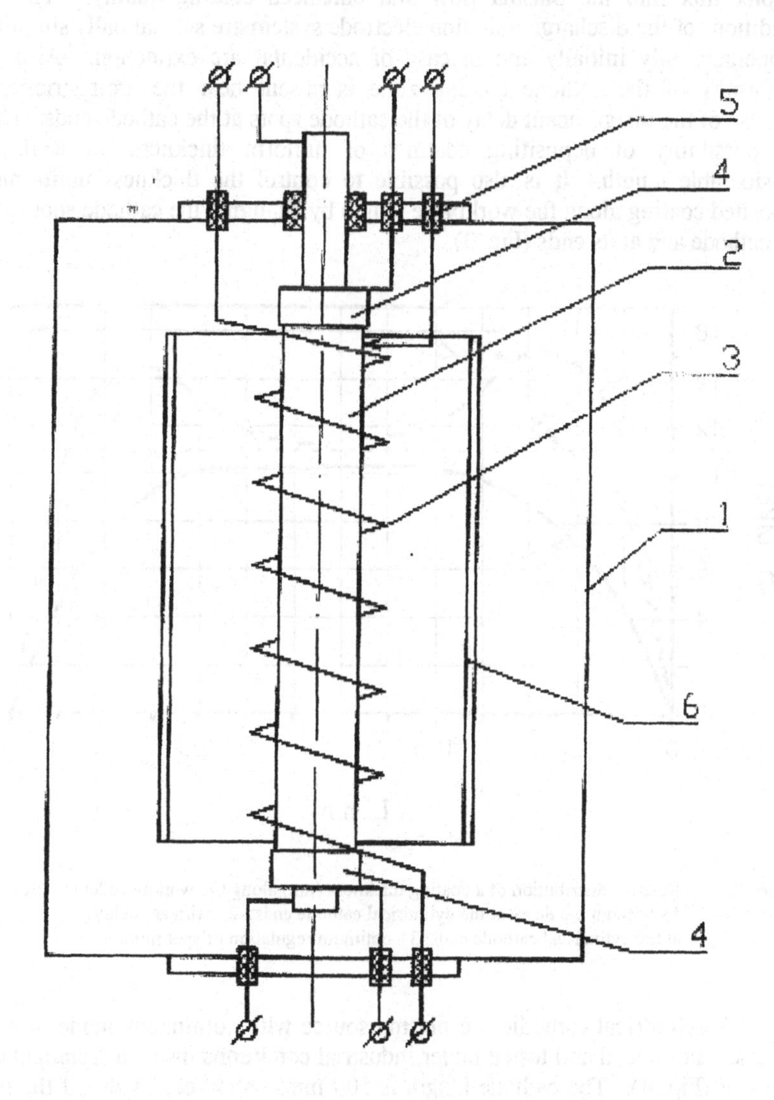

Figure 2. **Scheme of the long cylinder-type vacuum-arc evaporator: 1 - vacuum chamber; 2 - cathode; 3 - anode; 4 - arc-restrictive screens; 5 - vacuum lead-ins and inputs; 6 - item to be coated**

spots reach the arc restriction screens, and consequently the direction of cathode spot motion is changed to the reverse direction. Thus the steered cathode spots perform a rotary motion around the cylindrical cathode and reciprocate continuously along the cathode with controlled frequency and velocity. This vacuum-arc source provides the possibility of increased evaporator productivity. The uniform distribution of ion current density along the cathode axis and the ordered mode of the continuous cathode spot motion along the cathode working surface lead to a reduction in macroparticle droplet flux into the plasma flow and enhanced coating quality. The operation conditions of the discharge initiation electrode system are substantially simplified since it operates only initially and in case of accidental arc extinction. Also, the non-uniformity of the cathode erosion zone is absent near the arc-restriction screens because of the insignificant delay of the cathode spots at the cathode ends. This allows the possibility of depositing coatings of uniform thickness on work-pieces of considerable length. It is also possible to control the thickness uniformity of the deposited coating along the workpiece length by changing the cathode spot speed along the cathode and at its ends (Fig. 3).

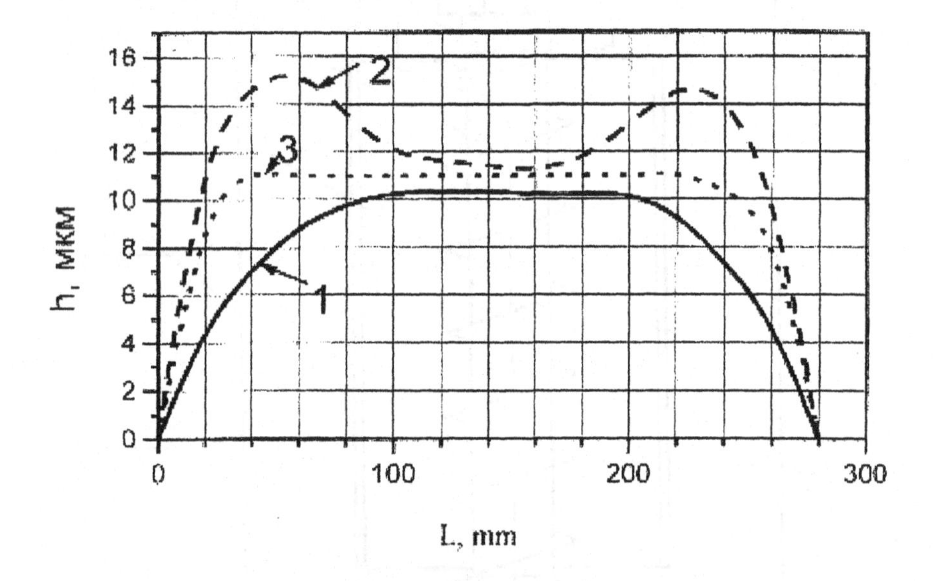

Figure 3. Relative distribution of a coating thickness (μm) along the workpiece length (mm). 1 - without arc delay at the cylindrical cathode ends; 2 - with arc delay at the cylindrical cathode ends; 3 - optimum regulation of spot motion.

A cylindrical cathodic arc plasma source with continuous mode of arcing was designed, fabricated and tested under industrial conditions inside a rectangular vacuum chamber (Fig. 4). The cathode length is 500 mm. An overall view of the batch-type installation is shown in Fig. 5. It is also equipped with a planetary system for parts to be coated, and with two (or three if required) butt-type vacuum-arc sources with coaxial electrode system for evaporation of required materials. This allows making comparative studies of the microstructure and properties of deposited layers for the different kinds of cathodic arc evaporators.

Figure 4 Linear cylindrical cathodic arc evaporator with lateral working surface, inside the vacuum chamber. The titanium cathode is of diameter of 100 mm and length 500 mm.

The volt-ampere characteristics were obtained for the case of full batch loading with long items: arc current 150 to 300A, arc voltage 25 to 40V, starting voltage 80 V. Substrate bias voltage was 150 V.

We also designed, fabricated and tested an electronics system for arc triggering by means of current conduction between electrodes on the insulator surface which are partly coated with a thin graphite layer and fully protected against metal deposition by a thin-wall metal tube. Ignition voltage was 5 kV and current 100 mA. The cathode material is evaporated and ionized at a breakdown zone between the thin-wall tube and the cathode. It was found that the direction and velocity of the cathode spots along the cylindrical cathode can be controlled by the current value and its commutation in the cathode circuit.

Figure 5 Overall view of the vacuum arc installation with the vertical cylindrical cathode at the center of the working chamber and two butt-type round cathodes on the lateral walls of the chamber.

4. Some Applications

The results obtained with the linear 500 mm long arc evaporator were used as a basis for the development of vacuum arc evaporators with tubular cathodes 32 mm and 100 mm in diameter and 1550 mm in length. These evaporators were used for deposition of functional coatings on the inner surfaces of long, round steel tubes. An overall view of the evaporator with a 1550 mm Ti cathode is shown in Fig. 6. Operational parameters of were arc current 300 to 400 A and arc voltage 30 to 45. A vacuum arc evaporator with a 1550 long and 32 mm diameter Ti cathode is shown in operation in Fig. 7. Comparison of the deposition rates for a butt-type Ti cathode with 68 mm ingot cathode diameter, and linear cylinder-type deposition systems with titanium cathodes 100 mm in diameter and 500, 1550 mm in length with a 180 mm substrate-to-cathode distance gave values of 16, 10 and 2 μm/hour.

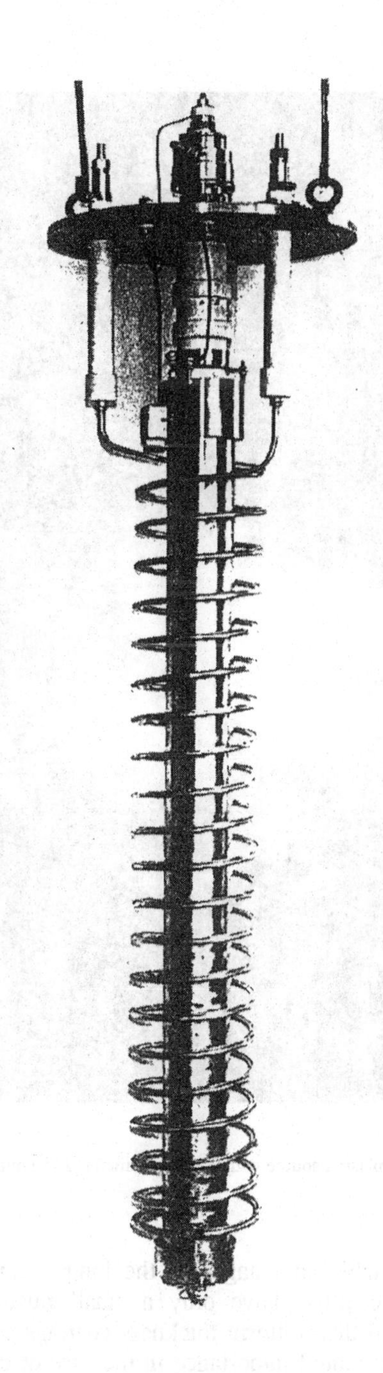

Figure 6 **Vacuum arc plasma source with titanium cathode 100 mm in diameter and length 1550 mm.**

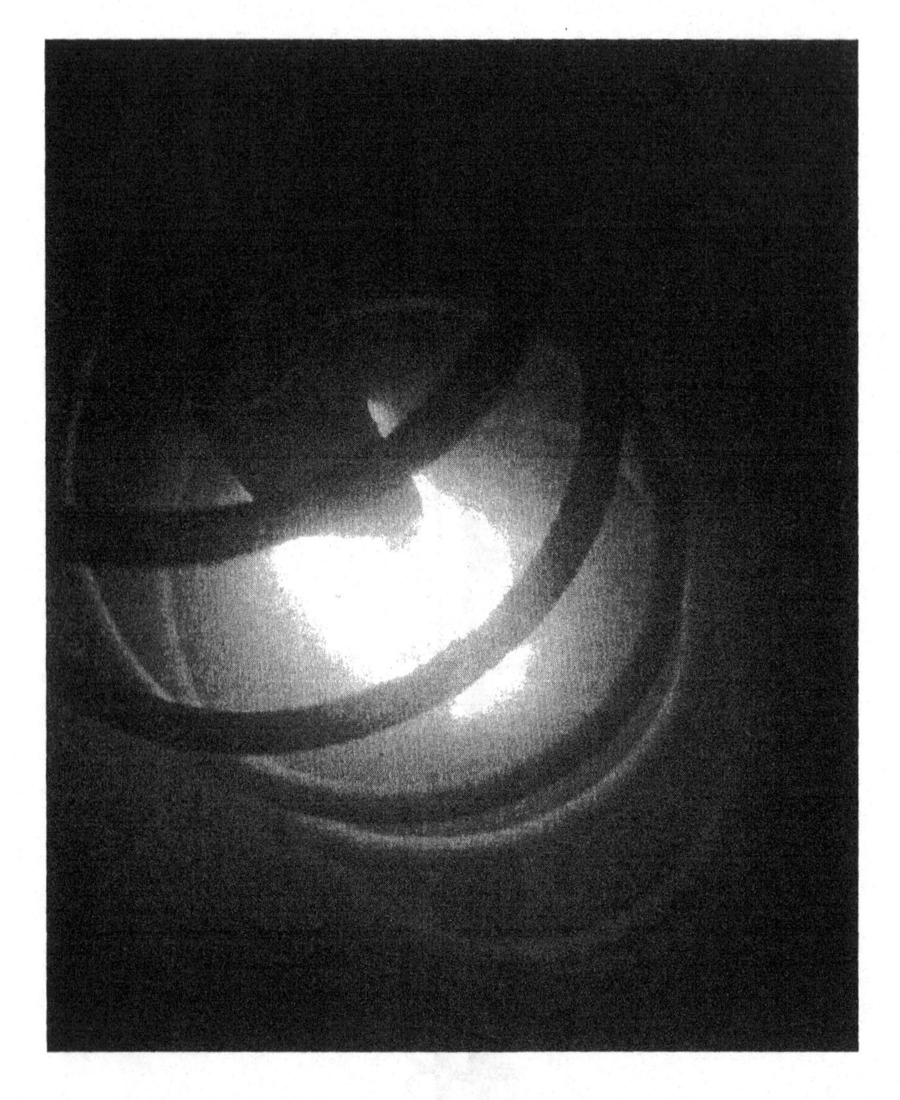

Figure 7 Linear vacuum arc plasma source with 32 mm diameter, 1550 mm long, titanium cathode, in operation

Among the indubitable advantages of the long vacuum arc evaporator with steered arc is that the coatings have only a small microdroplet (macroparticle) contamination (Fig. 8) and that uniform thickness coatings can be deposited on long items. This can be of paramount importance in the case of treatment of such critical and expensive products as laser graved anilox rolls for the printing industry, various shafts and critical machine parts for the textile, wood-pulp and paper industries, and for the for machine-building and electric-power industries.

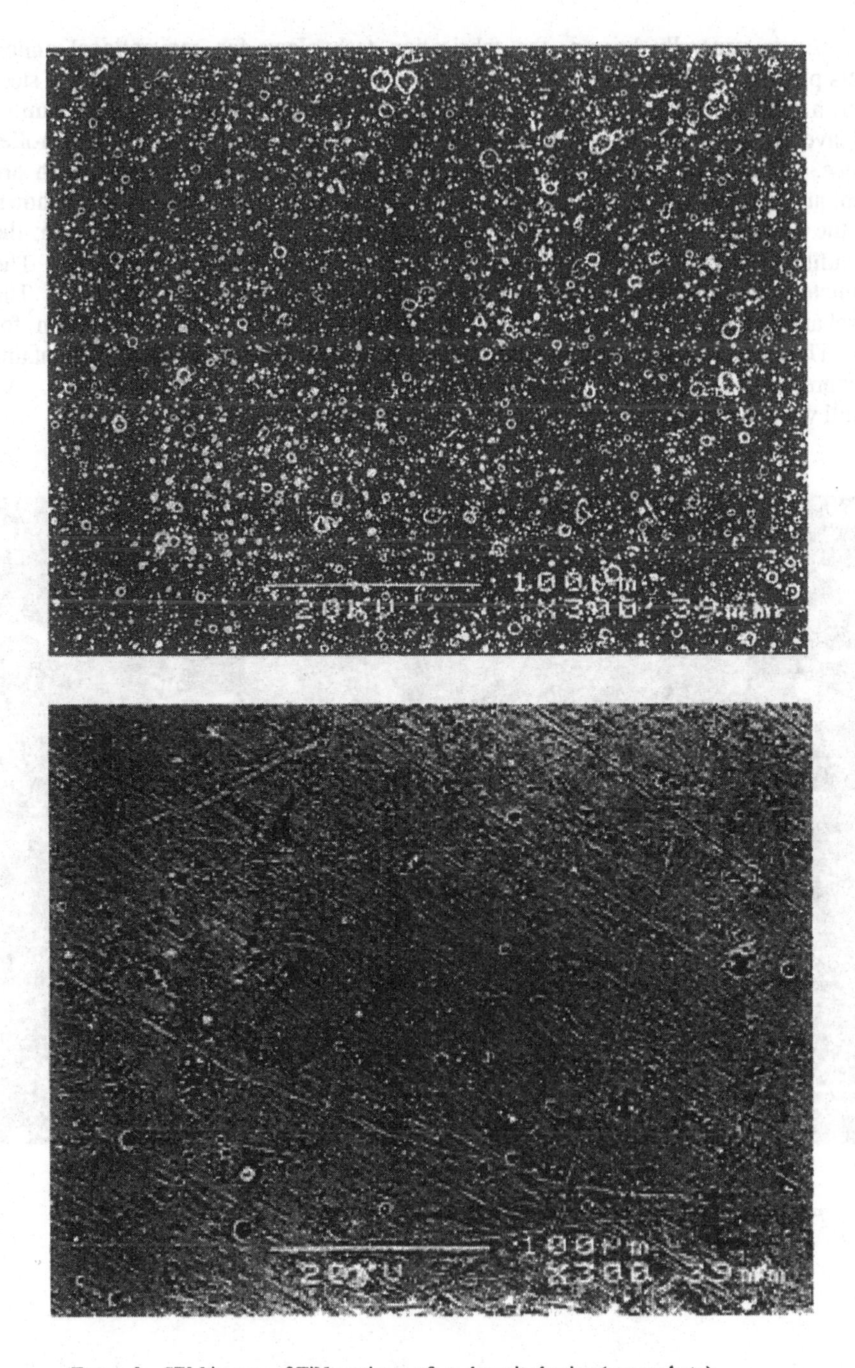

Figure 8 SEM images of TiN coating surface deposited using (upper photo) random mode, and (lower photo) steered arc mode.

As is generally known, the fabrication technology for conventional anilox rollers provides the deposition of two-layer coatings onto the external surfaces of steel roller: a copper layer, engraved by means of diamond tools, and a hard chromium or TiN layer in order to increase the wear resistance of the cellular structure of the roller surface. Studies of two-layered Cu/TiN coatings deposited by steered vacuum arc techniques onto prototype cylindrical steel rollers with a ground surface have shown that the microstructure of Cu and TiN condensates is columnar in cross-section; the crystallite average widths are 1 to 3 μm and 0.1 to 0.2 μm, respectively. The parameters of the crystal lattice are 3.603 Å for Cu and 4.26 Å for TiN. The microhardness of the deposited layers is 85 kg/mm^2 for Cu and 1900 - 2100 kg/mm^2 for TiN. The surface microstructure of the TiN (Fig. 8 b) and Cu layers (Fig. 9) contains no appreciable defects and the coating has good interlayer-to-substrate adhesion. An overall view of a prototype roller with the two-layered coating is shown in Fig. 10.

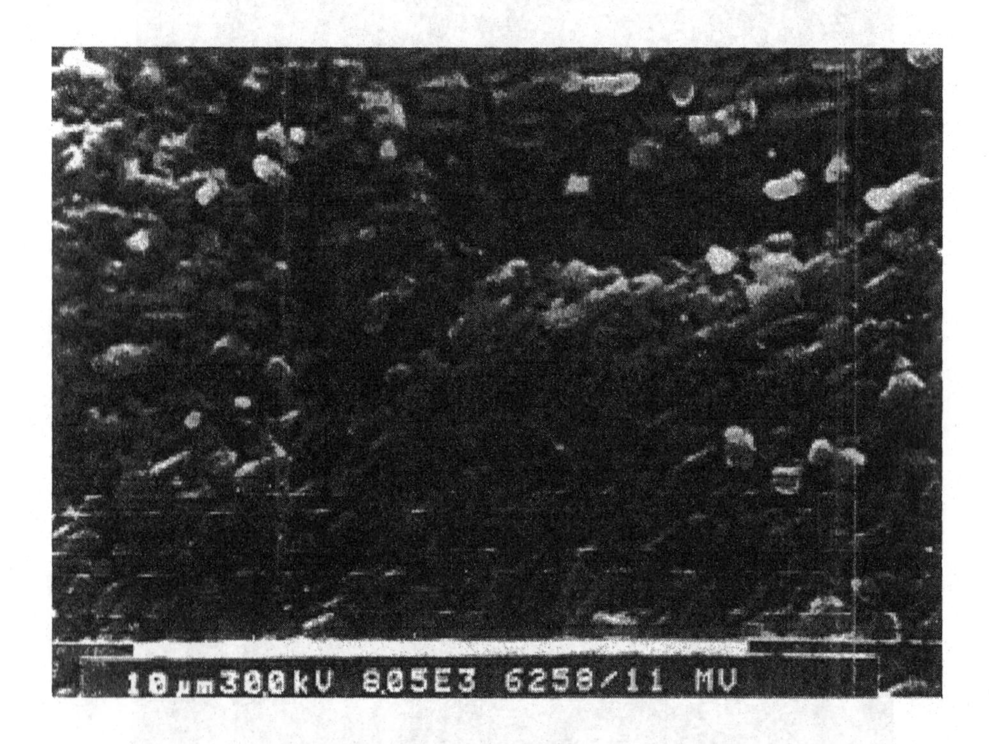

Figure 9 SEM image of Cu coating surface deposited using steered arc mode

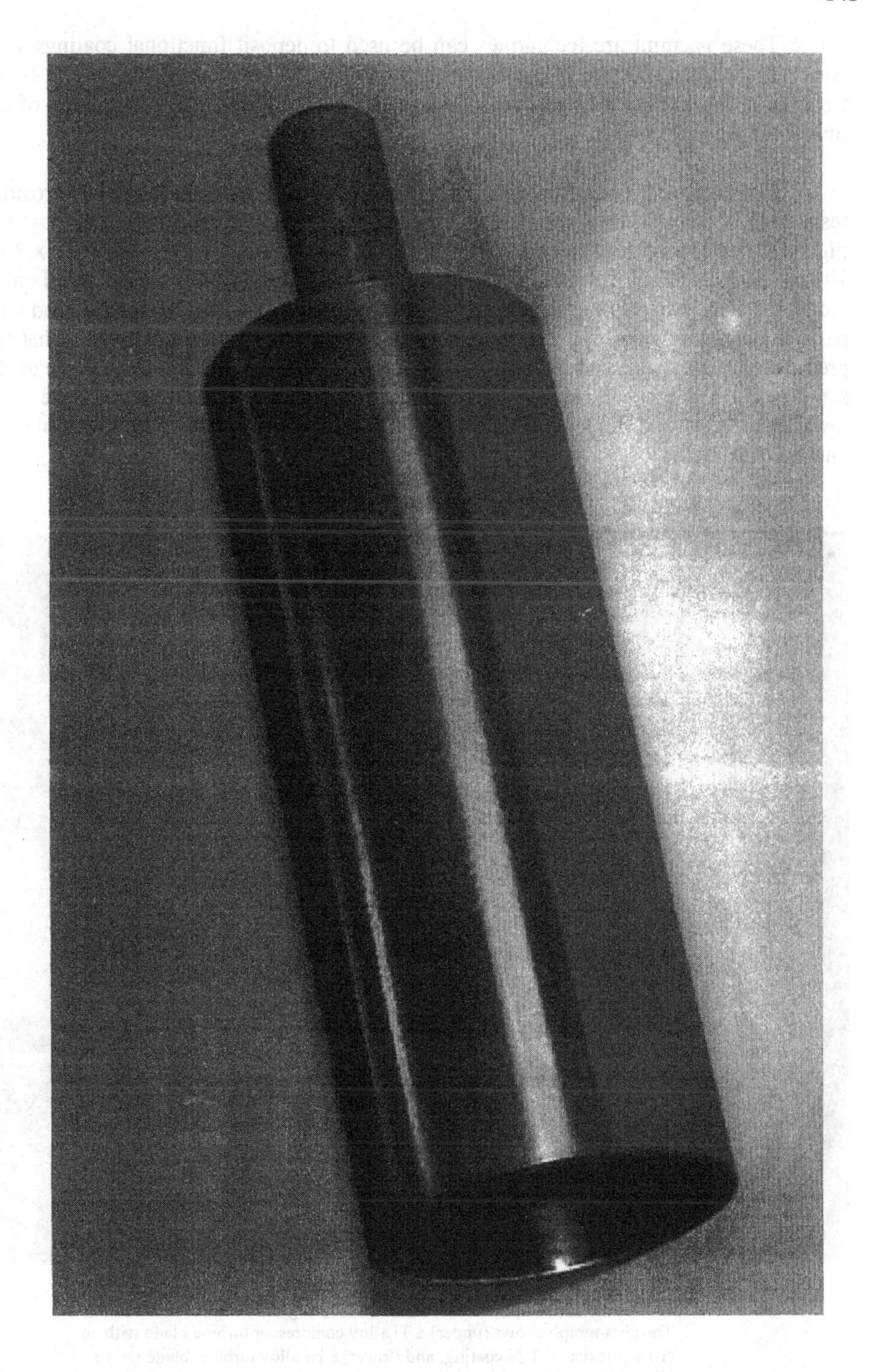

Figure 10 **Prototype steel roller with two-layered Cu/TiN coating. Diameter is 100 mm and length is 250 mm.**

These vacuum arc techniques can be used to deposit functional coatings with good adhesion irrespective of the roller material (steel or aluminum alloys, e.g. B95). Good adhesion in this case is of great importance since it determines the quality of the final laser graved product.

Vacuum arc evaporation with steered arc has also been used to deposit erosion-resistant TiN coatings onto compressor turbine blades made of BT6 titanium alloy; see Fig. 11. The TiN coating thickness was 8 μm. Wear resistance tests of flat 70 x 30 x 10 mm samples made of the same material, with and without a TiN, were carried out at the ICA-1 installation under conditions of wear by free abrasive sand (quartz sand with particle diameter in the range 0.05 to 0.4 mm). The test results have shown that the presence of the vacuum arc deposited hard coating increased the wear resistance of the samples by a factor of three. Test conditions were: test rod rotation rate - 25 revolutions/min; and for the holder of fixed samples - 100 revolutions/min; test duration - 10 hours.

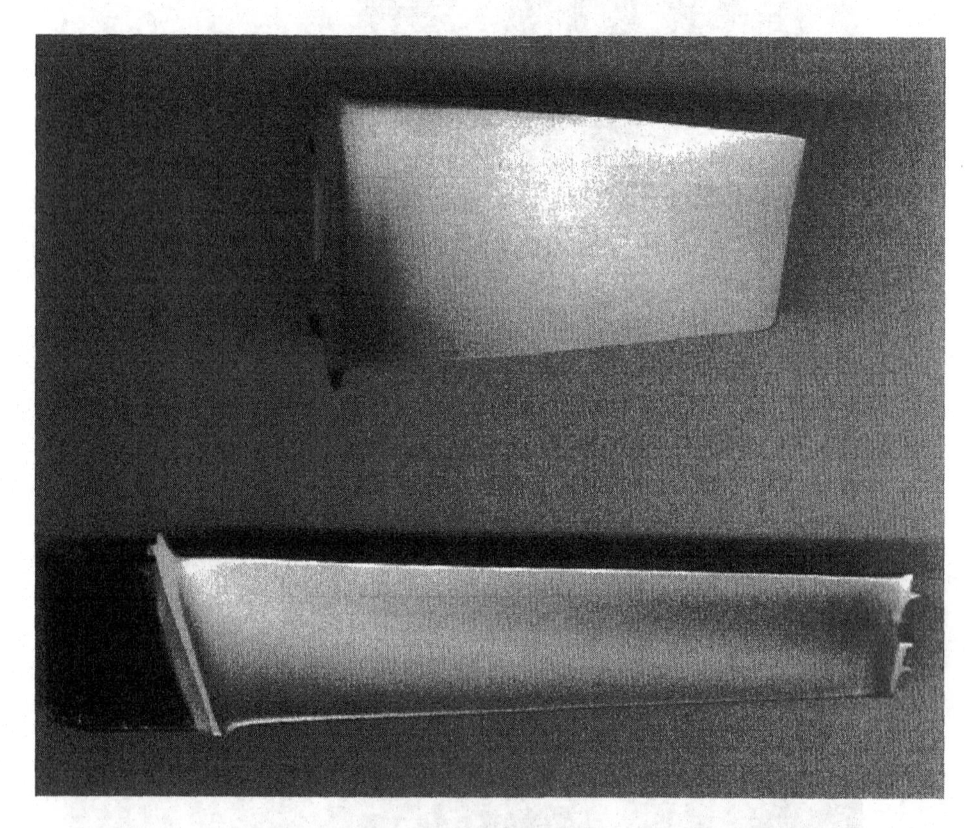

Figure 11 Ti alloy and Ni alloy turbine blades coated using the vacuum arc method. The photograph shows (upper) a Ti alloy compressor turbine blade with an erosion resistant TiN coating, and (lower) a Ni alloy turbine blade with a NiCrAlY bond coat.

As is generally known, heat-resistant alloys of the Me-Cr-Al-Y system, where Me is Ni, Co or Fe, are widely used for protection of gas turbine blades and vanes against gas corrosion and high temperature erosion [8]. At present, electron beam physical vapor deposition and plasma coating in a protective atmosphere are the main techniques used in this field. The application of vacuum arc technology for this purpose is of great interest since a substantial ion component (20 to 40%) is present in the vapor flow in this case. This leads to significant improvement in deposition conditions and provides good adhesion at lower deposition temperature.

A 50 μm thick vacuum-arc-deposited NiCrAlY coating (20%Cr, 9%Al, 0.35%Y, Ni base) was deposited onto cylindrical samples and blades made of ЖС6У alloy (Fig. 11). The samples were investigated by scanning electron microscopy (SEM). X-ray photoelectron spectroscopy / electron spectroscopy for chemical analysis (XPS/ ESCA) was used to determine the chemical composition of the coating. SEM surface and cross-sectional images of NiCrAlY coating deposited with a steered arc mode are shown in Fig. 12. The results of the investigations have shown a uniform distribution of alloy elements in the coating and the absence of pores at the substrate-coating interface. Heat-resistance tests of the samples and blades, with and without coating, at 1000°C heating temperature in open air for 20, 50, 100 and 200 hours, showed that the coatings did not failed, did not spall from the substrates, and the coated items have much better heat-resistance compared to the uncoated samples.

The vacuum arc deposition technique was also used for "healing" of casting defects (micropores, pittings) on blade and vane surfaces. In this case an alloy of the NiCrSi system (66% Ni, 24% Cr, 10% Si) with melting temperature 1095°C was used as the coating material. The coating thickness was a function of defect size and the average value was 10μm. Coated samples were then placed in a heating vacuum furnace in order to fuse the deposited layer. A surface liquid film filled in all surface microdefects on the blades and vanes.

The use of metallic alloys with different chemical compositions as cathode materials for vacuum arc evaporation substantially widens the application fields of the method. For example, the corrosion resistant X45Y4 alloy of the FeCrAl system was used for coating deposition onto plunger type steel samples so as to increase their corrosion- and abrasion-resistance to the action of molten glass. Testing of coated samples against pressure with molten glass showed:
- the absence of coating spalling from the substrate after thermal cycling, reflecting good adhesion between the coating and the substrate;
- high oxidation-resistance of the coating owing to the presence of a high quantity of Cr in the coating;
- a 60% increase in life-time of 10 μm coated punch-type samples in comparison with uncoated samples after pressure against molten glass.

These results are encouraging for further investigations in this area.

The metallization of fiber-woven roll materials opens up the possibility of modifying their properties. The surface structure of a graphite cloth with Al coating is shown in Fig. 13.

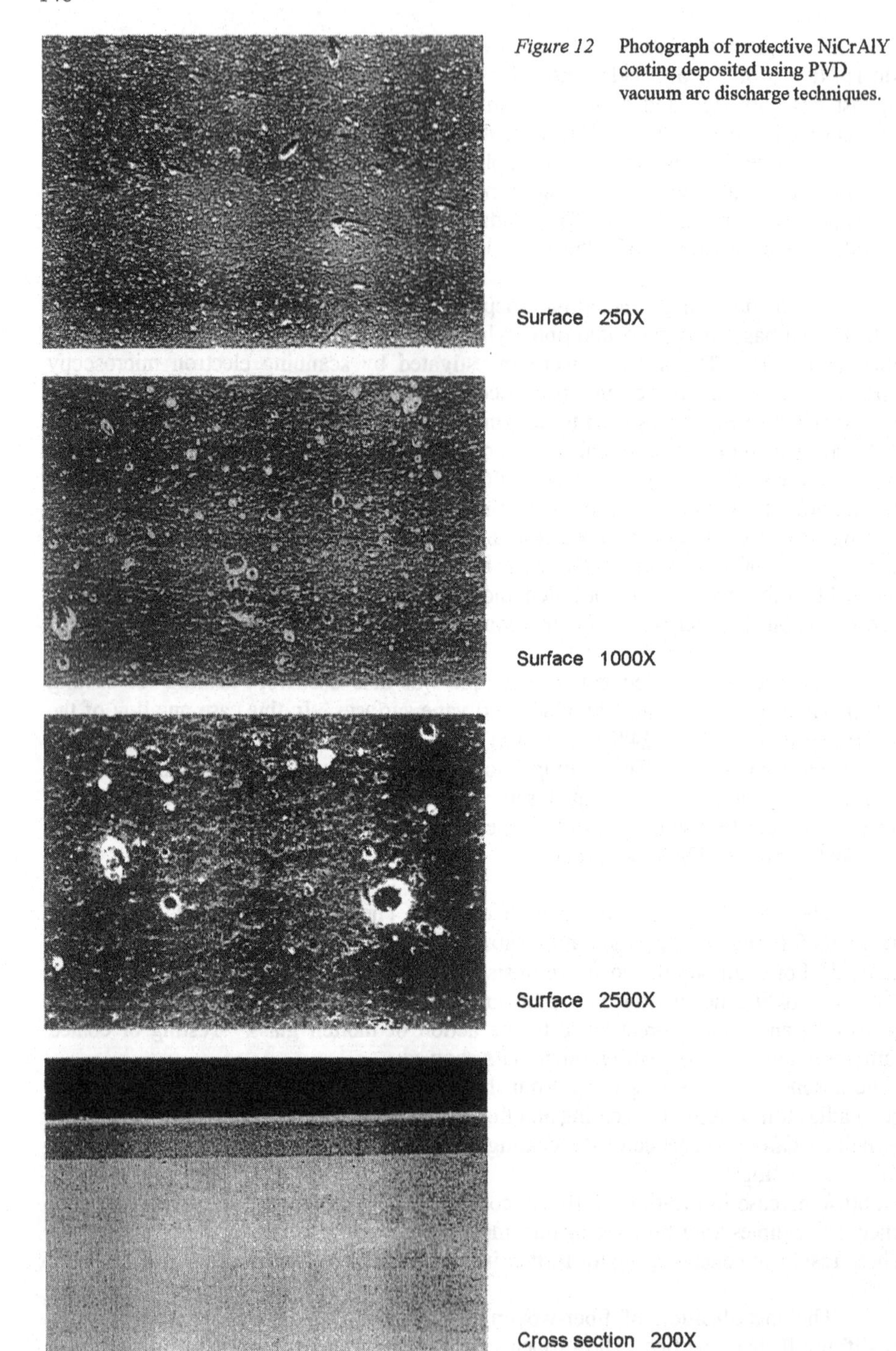

Figure 12 Photograph of protective NiCrAlY coating deposited using PVD vacuum arc discharge techniques.

Figure 13 **Surface of woven graphite fabric that has been coated with aluminum**

15X (picture width ~9 mm)

1000X (picture width ~135 µm)

2000X (picture width ~65 µm)

8000X (picture width ~16 µm)

The electrical conductivity of the Al-coated woven carbon fabric increases with increasing coating thickness. The surface resistance decreases from 5 - 10 ohms per square at 10 μm Al coating thickness and is essentially zero at 15 – 18 μm Al coating thickness. This technique was also used to deposit a Cu coating onto carbon fabric in order to improve its antiseptic and antimicrobial properties. As a result of this work a vacuum arc installation was designed with a vacuum chamber diameter of 1.0 m and height of 1.4 m.

5. Conclusion

A new kind of long, linear cylinder-type vacuum arc evaporator with continuous mode of operation has been designed, fabricated and tested, and a number of examples of its application have been described here. The length of the cathodic arc plasma source is up to 1550 mm, and the device has an external side wall working surface. Functional coatings of pure metals, multi-component alloys, and refractory compounds on long and large items have been deposited. Among the applications that have been investigated are: layered Cu/TiN coatings deposited by steered vacuum arc techniques onto prototype cylindrical steel rollers; erosion-resistant TiN coatings on compressor turbine blades made of BT6 titanium alloy; NiCrAlY coatings on turbine blades made of ЖС6У alloy; corrosion resistant X45Y4 FeCrAl alloy on steel substrates to increase their corrosion- and abrasion-resistance against molten glass; and the metallization of woven carbon-fiber fabric material. In all of these cases the results were excellent. The advanced cathodic arc deposition systems that we have developed provide the possibility of increasing significantly the working characteristics of critical machinery parts in modern engineering industry.

References

1. Tolok V.T., Padalka V.G. (1979), "Development and implementation of new methods for high-energy vacuum-plasma technology", *Herald of the Ukrainian Academy of Sciences*, No.4, 40-50.
2. Popov V.F., Gorin Yu.N. (1988), *Processes and Installations of Electron-Ion Technology*, 256 p. High-School Publishers, Moscow.
3. Dorodnov A.M. (1978), "Technological plasma accelerator", *J. Technical Physics* **48**, No.9, 1858-1870.
4. Saksaganskiy G.L. (1988), *Electrophysical vacuum pumps*, 277 p. Energoatom Publishers, Moscow.
5. Vetrov N.Z., Lisenkov A.A. (2001), "Lengthy-type vacuum-arc sources", in Proceedings of the 4th International Symposium on "Vacuum Technologies and Equipment", Kharkiv Institute of Physics and Technology (KIPT), Part III, 339-342, Kharkiv, Ukraine.
6. Abramov I.S., Bystrov Yu.A., Lisenkov A.A. (1997), *A vacuum-arc plasma source*, Patent No.2072642, Russia, MKI C23c 14 / 32, Bull. Of Inventions No.11 (1997) Moscow.
7. Abramov I.S., Bystrov Yu.A., Lisenkov A.A. (1997), *A vacuum-arc plasma source*, Patent No. 2098512, Russia, MKI C23c 14 / 32, Bull. Of Inventions No. 34 (1997) Moscow.
8. Movchan B.A., Malashenko I.S. (1983), *Vacuum-deposited heat-resistant coatings*, "Naukova Dumka" Publishing House, 232 p., Kyiv.

9. Viazovikina N.V., Zhalenko N.A., Kurapov Yu.A. et al. (1991), "Influence of Cr and Impurities on Electrochemical and Corrosive Properties of FeCrAl System Alloys", *J. Metal Protection*, **27**, No.6, 911-916.
10. Strikovsky A.V., Kostrov A.V., Gundorin V.I., Ovechkin M.M., Chudner R.V. (2000), *The properties of Aluminium covered by method of vacuum-arc deposition.* In Proceedings of the 5th Conference on Modification of Materials with Particle Beams and Plasma Flows, **3**. Ed. by G.Mesyts and A. Ryabchikov. (2000) Tomsk, Russia

ARC GENERATORS OF LOW-TEMPERATURE PLASMA AND THEIR APPLICATIONS

N. N. Koval and P. M. Schanin

High Current Electronics Institute
Russian Academy of Sciences
4 Akademichesky ave., Tomsk
634055, Russia

Abstract. In this paper, three types of arc plasma generators with a hollow cathode are presented in which the gas arc discharge is initiated by electrons emitted by a hot filament, or with the use of a trigger system based on a surface dielectric discharge, or by an additional glow discharge. The generators produce gas plasmas of densities 10^{10} – 10^{12} cm^{-3} in large volumes of up to 1 m^3 at discharge currents of 100–200 A and low pressure of 10^{-1}–10^{-2} Pa. Consideration is given to some operating features of the plasma generators and to their use for investigating the surface modification of solid materials, including steel nitriding and plasma-assisted deposition of TiN and *a*-C:H films and in the development of plasma emitters for high-current electron sources.

1. Introduction

The low operating voltage, the high discharge current, and the wide range of pressures at which discharges show stable initiation and burning have stimulated the extended study of such discharges and their application in various devices.

They are used in high-current charged-particle sources [1–4] and plasma generators [5, 6] which provide plasma densities of 10^{10}–10^{13} cm^{-3} at low pressures. In electron sources with plasma emitters, due to their high emissivity, the use of pulsed discharges allows one to produce high current densities of up to 100 A/cm^2 at microsecond pulse durations and low pressures of 10^{-1}–10^{-3} Pa. The acceleration gap has a sufficiently high electric strength at these pressures. In comparison with hot cathodes, plasma emitters are insensitive to ion bombardment and to increasing pressure during their operation and show a higher energy efficiency in pulsed operation of the electron source.

The energy efficiency and productivity play an important role in the technological processes of ion nitriding and plasma-assisted deposition of coatings at high continuous discharge currents. Continuous arc discharges, at currents of up to 200 A, allow one to produce homogeneous and high-density gas plasmas in volumes of several cubic meters. In combination with the wide range of operating pressures at which plasma is produced, such discharges make it possible to produce various types of ion-plasma surface modification of materials: hydrogen-free ion nitriding [7, 8], cleaning and activation of surfaces before deposition of hard and corrosion-resistant coatings, and low-energy ion implantation [9]. The gas plasma of a chemically active gas, such as O_2 or CH_4, is produced when a cathode spot operates on a cold surface.

E. Oks and I. Brown (eds.),
Emerging Applications of Vacuum-Arc-Produced Plasma, Ion and Electron Beams, 151–162.

1. The plasma generator with a hot and a hollow cathode

The design of a plasma generator with a cylindrical hollow cathode is shown in Fig. 1. Cylindrical hollow stainless-steel cathode 4 of internal diameter 90 mm and length 280 mm is mounted on water-cooled flange *2*. Hot cathode *5* made of tungsten wire of diameter 1.8 mm is located on two copper water-cooled current leads *1* inside the cavity. A longitudinal magnetic field of induction $2.5\cdot10^{-2}$ T, which stabilizes the discharge, is induced by solenoid *6*. The cathode unit is fixed to water-cooled case *8* through insulating gasket *3*. The case *8* of the plasma generator is at the anode potential and is connected to the vacuum chamber of dimensions 600×600×600 mm. In this discharge system, the vacuum chamber plays the role of a hollow anode. The plasma generator is powered from a filament-circuit transformer which provides a current of up to 180 A at a voltage of 12 V and from a three-phase rectifier allowing gradual current tuning in the range from 10 to 180 A at an open-circuit voltage of 70 V. The two-channel automatic gas leak-in system ensures gas leak-in.

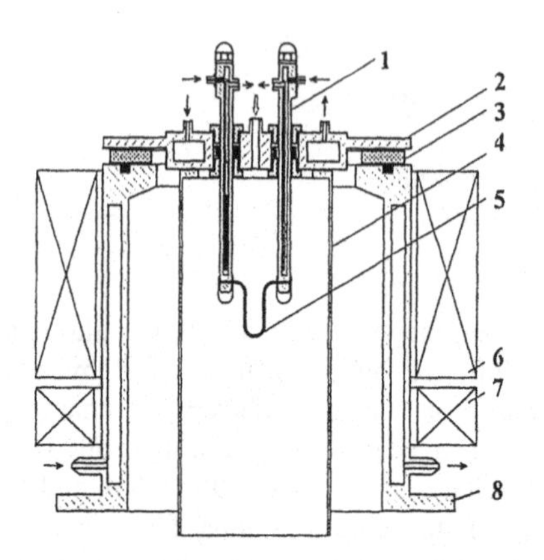

Fig. 1. Plasma generator with a cylindrical hollow cathode: *1* – water-cooled current leads; *2*- water-cooled flange; *3* – insulator; *4* – cylindrical hollow cathode; *5* – hot cathode; *6* – solenoid; *7* – focusing coil; *8* – case.

Since the path of the electrons emitted by the hot cathode increases in the magnetic field, these electrons efficiently ionize the working gas and generate gas-discharge plasma in the hollow cathode. The unmagnetized ions accelerated in the region of near-cathode potential fall knock secondary electrons out of the walls of the hollow cathode. The processes of ionization in the cavity are intensified, and the conditions are thus established for the initiation and operation of a high-current gas discharge at low discharge pressures and voltages. By varying the filament current and, consequently, the electron emission from the cathode it is possible to control the discharge current and to initiate a non-self-sustained arc discharge without a cathode spot. The main characteristics of the discharge have been obtained under the following experimental conditions: the hot cathode was connected to the cavity and the working gas was argon.

Figure 2 shows the current-voltage characteristic of the discharge at a filament current I_f = 135 A and at two pressures. Figure 3,*a* depicts the discharge operating voltage U_d and the discharge current I_d versus the filament current I_f at a pressure $p = 3\cdot10^{-1}$ Pa. The pressure dependences of the discharge operating voltage and discharge current presented in Fig. 3,*b* are similar in character, i.e., as the pressure is increased in the range from $7\cdot10^{-2}$ to $9\cdot10^{-1}$ Pa, the discharge current increases, while the discharge operating voltage decreases.

Further increasing the length of the hollow cathode causes a rapid increase in the pressure of stable initiation of the discharge, because it is difficult for the cavity to take the anode potential, whereas decreasing the length of the hollow cathode provides a decrease in discharge current.

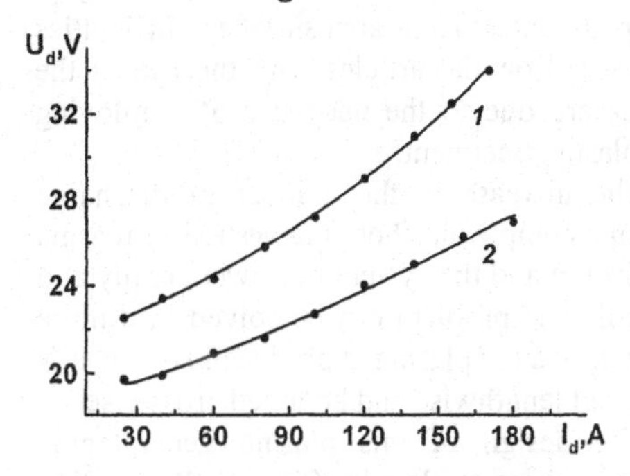

Fig. 2. Current-voltage characteristics of the discharge at two pressures p, Pa: *1* - $3{\cdot}10^{-1}$; *2* - $5{\cdot}10^{-1}$.

In non-self-sustained arcing, the hollow cathode plays an important role. The data of Table 1 show how the length of the hollow cathode affects the characteristics of the discharge at a constant cavity diameter $D = 9$ cm. In this table, L is the length of the hollow cathode, I_d is the discharge current, U_d is the operating voltage, and p_{in} is the pressure at which a non-self-sustained arc discharge is initiated. As shown in [6], the optimum ratio between the length and the diameter of the hollow cathode, L/D, is 3–4.

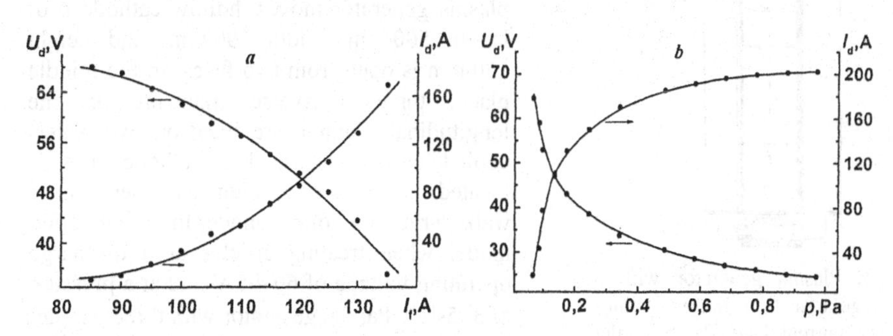

Fig. 3. Discharge operating voltage and discharge current versus the filament current (*a*) and pressure (*b*).

Table 1.

L, cm	I_d, A	U_d, V	p_{in}, Pa
0	15	74	$1{\cdot}10^{-2}$
28,5	40	60	$7{\cdot}10^{-2}$
31,0	42	59,5	$9{\cdot}10^{-2}$
34,0	47	59	$16{\cdot}10^{-2}$

At discharge currents $I_d = 90$–150 A, operating voltages $U_d = 60$–70 V, pressures $p = 1{\cdot}10^{-1}$ Pa, the generator produces nitrogen and argon plasmas of density 10^{10} cm^{-3} and with a spatial distribution homogeneous within ±20%. The plasma has a positive potential of the order of +10 V relative to the anode. The electron temperature is $T_e = 4$ eV. Probe measurements have revealed that there also exists a group of fast electrons with temperature $T_e = 10$ eV in the plasma. With a negative bias applied to the probe, the saturation ion current is 5 mA/cm^2.

2. The plasma generator with a rectangular-section hollow cathode

With large-dimension setups for plasma treatment of large-area surfaces, difficulties emerge, particularly, when coatings are deposited on flat articles with the use of the discharge discussed above. These difficulties are due to the necessity of employing several discharge systems to ensure uniform plasma treatment.

The increase in the number of discharge systems complicates both the system of plasma generation and the system of power supply and control. The problem can be solved in a more efficient way if plasma is produced so that it is extended lengthwise and bounded crosswise.

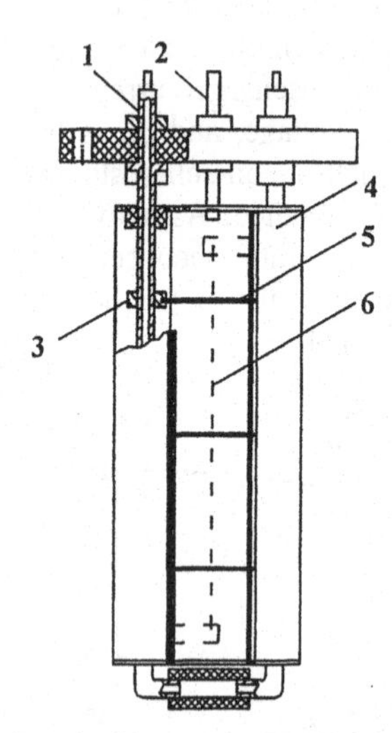

Fig. 4. Plasma generator with a rectangular-section hollow cathode: *1* - current lead, *2* – gas inlet pipe; *3* - filament holder; *4* – hollow cathode; *5* – transverse filament; *6* – longitudinal filament.

The design of this plasma generator is shown schematically in Fig. 4. Proceeding from practical requirements of simultaneous treatment of two flat articles between which the plasma generator moves, hollow cathode *6* of length 500 mm, width 100 mm, and height 150 mm is open from two faces. In the middle plane, three transverse filaments or one longitudinal filament are fixed on two water-cooled current leads. The cathode unit is located inside the vacuum chamber whose walls serve as a hollow anode. In the operating mode, when treating articles at a discharge operating voltage of 60–65 V and at a pressure of $3–5\cdot10^{-1}$ Pa, the generator with three parallel filaments or one longitudinal filament provides a discharge current of 20–30 A at a total filament current of 150 A or at a filament current of 50 A, respectively. Investigations have shown that if only one face of the hollow cathode is open, the plasma density increases by a factor of 1.5. The plasma density measured at a distance of 100 mm from the hollow cathode, n_e, is 10^{10} cm^{-3}, the electron temperature, T_e, is 4 eV, and the maximum density of the saturation ion current, j_i, is 5 mA/cm^2. The density distribution of the saturation ion current along the longitudinal axis of the hollow cathode is given in Fig. 5. As a result of the studies conducted, a full-scale plasma generator comprising three identical series-connected elements with a longitudinal arrangement of the filaments has been designed. With parallel power supply of the filaments of total filament current 120 A,

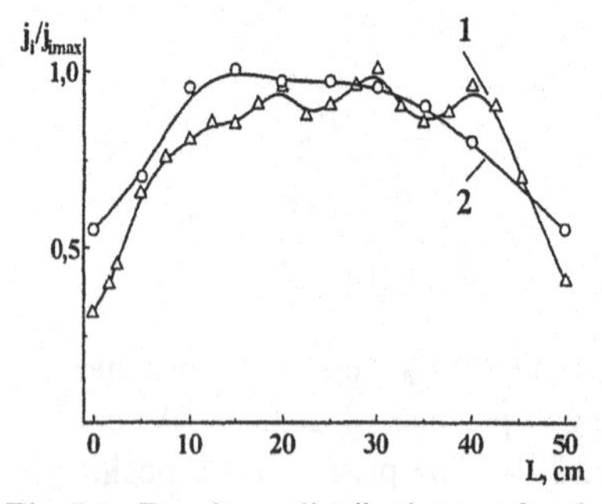

Fig. 5. Density distribution of the saturation ion current along the longitudinal axis of the system: *1* - transverse filaments; *2* – longitudinal filament.

the plasma generator produced a discharge current over 50 A in the operating mode. Due to flashover in the plasma at the face edges of some units, the uniformity of the plasma density distribution along the longitudinal axis of the hollow cathode in the 1400 mm long working irradiation zone was within ±10%.

The designed plasma generator was employed in the technological process of finish cleaning of architectural glasses before deposition of functional coatings.

3. The plasma generator with a cold hollow cathode

The lifetime of a plasma generator with a hot cathode, when it operates in an active gas (O_2, CH_4), is limited to about several hours because of the oxidation of the hot filament and its destruction by ions.

In this design, hollow cathode *1* (Fig. 6) of diameter 100 mm and length 150 mm made of stainless steel is completely cooled with water. The arc breaker *4* performed like a tumbler is fixed on the face of the cathode through insulator *6* and is at a floating potential. The arc breaker prevents arcing over the insulator surface and serves as a barrier for the penetration of droplets into hollow anode (vacuum chamber) *7*. The discharge is initiated at an operating pressure ranging from $1 \cdot 10^{-2}$ to $1 \cdot 10^{-1}$ Pa when a voltage of the order of several kilovolts is applied to trigger *2*. Short magnetic coil *3* creates an axial magnetic field.

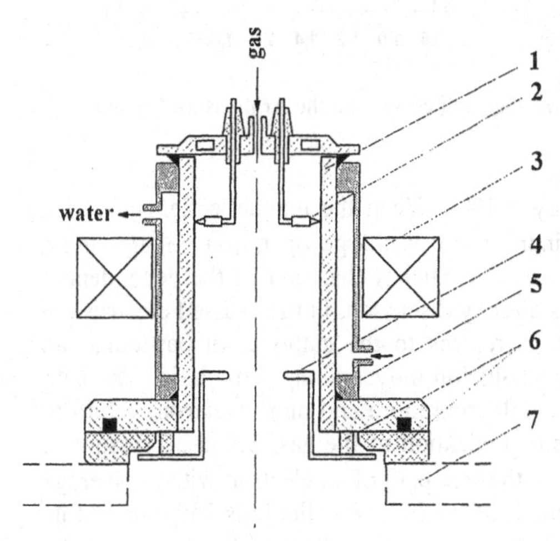

Fig. 6. Plasma generator with cold hollow cathode.

In the range from 10 to 120 A and with the gas pressure and the magnetic field kept constant, the discharge voltage remains constant, i.e., the plasma generator has a plane current-voltage characteristic. The gas pressure in the hollow cathode and the magnetic field have the most profound effect on the discharge characteristics (Fig. 7).

The electrons accelerated in the cathode fall potential region ionize the working gas leaked in the cavity and initiate a self-sustained arc discharge. In the crossed electric (between the cathode and the plasma) and axial magnetic fields, the cathode spot executes a rotation. The velocity of the spot increases with magnetic field.

Some of the electrons participate in the ionization of the gas, while the other, which do not experience inelastic collisions, return to the cathode following a complex path. The metal ions that have not experienced magnetization owing to the weak magnetic field, as well as the neutral atoms and the droplets generated by the cathode spot, find themselves on the opposite side of the cathode. The ions knock out secondary electrons, thus promoting stable operation of the discharge. Because of the ions, atoms, and droplet fraction, the material of the hollow cathode turns out to be oversputtered, and this increases its lifetime.

At a magnetic induction $B = 1.4 \cdot 10^{-2}$ T, an increase in pressure by an order of magnitude causes the discharge current to increase approximately two times and the

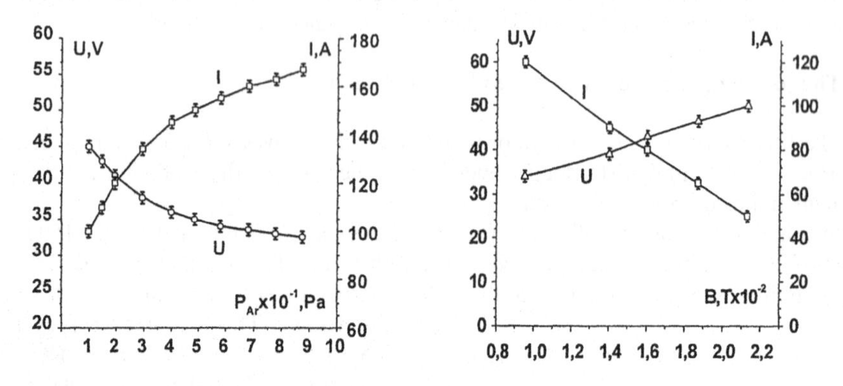

Fig. 7. Dependences of the discharge voltage and current on the Ar pressure (*a*) and the magnetic field (*b*).

discharge operation voltage to decrease by ~ 30%. When the magnetic field is varied from $1 \cdot 10^{-2}$ to $2 \cdot 10^{-2}$ T, a nearly linear increase in discharge operation voltage and a linear decrease in discharge current are observed. Such a character of the dependences can be explained in the following way. As already discussed, in the crossed electric and magnetic fields, some part of the electrons returns to the cathode. In particular, an electron which has not experienced even a collision moves along a trochoid. The path length decreases with increasing magnetic field, and more and more electrons return to the cathode, taking no part in the process of ionization of the gas. So, at a pressure of $1 \cdot 10^{-1}$ Pa, the trochoid length is smaller than the free path of an electron with an average energy of 20 eV. In the case where the length of the hollow cathode is 150 mm and its diameter is 100 mm, the discharge extinguishes at a pressure of $9 \cdot 10^{-3}$ Pa. Increasing the length of the hollow cathode causes an increase in minimum operating pressure and some increase in operating voltage.

In practice, in plasma-assisted deposition of functional coatings, an important factor limiting the range of application of electric-arc technologies is the uncontrollable ingress of the cathode material in the form of droplets from the plasma generator to the article under treatment. Table 2 compares the number of droplets coming into the chamber from the plasma generator and from a standard electric-arc evaporator in a time of 10 min.

Table 2.

Droplet diameter	Generator, mm^{-2}	Evaporator, mm^{-2}
up to 1 μm	4	$4 \cdot 10^4$
up to 2 μm	3	$5.5 \cdot 10^3$
over 2 μm	3	$2.0 \cdot 10^3$

At a discharge current of 100 A and Ar pressure $p = 5 \cdot 10^{-1}$ Pa, the generator produces plasma of density $n_e = 10^{11}$ cm^{-3} and electron temperature $T_e = 4$ eV in the chamber. The plasma has a positive potential of +4.5 V relative to the anode.

4. The pulsed plasma generator based on a glow-discharge-initiated arc

The discharge system is shown schematically in Fig. 8. The overall gas-discharge system consists of two discharge systems. The first one, formed by electrodes *1* and *3*, is intended for the initiation of an auxiliary discharge operating at a high pressure with the initiating pulse duration equal to 2–3 μs. The hollow cathode *1* is made as a cylinder of internal diameter 10 mm and length 50 mm. The second arc discharge is initiated by a pulse of duration up to 25 μs between the Mn ring insert installed inside the cylindrical electrode *2* of diameter 50 mm and length 40 mm, being the cathode for this discharge, and the anode consisting of two electrodes: cylindrical electrode *5* of diameter 70 mm and length 50 mm and grid electrode *6* with 0.3×0.3 mm meshes, placed on the end of the former one. To create a pressure difference with the aim to reduce the pressure in the cylindrical cathode *3*, the electrodes *3* and *5* are coupled through a hole, whose geometry was varied in experiments.

To reduce the igniting voltage of the initiating discharge, the cathode *1* is placed in a magnetic field of induction 0.1 T created by permanent ring magnets, and to promote the ignition of the glow discharge onto the grid anode *6*, the cylindrical anode is connected to the charging voltage source through a resistor of $R_2 = 100\ \Omega$. To ignite the initiating discharge, a pulsed voltage of amplitude up to 10 kV is applied from the secondary winding of a pulse transformer as a storage capacitor is switched on.

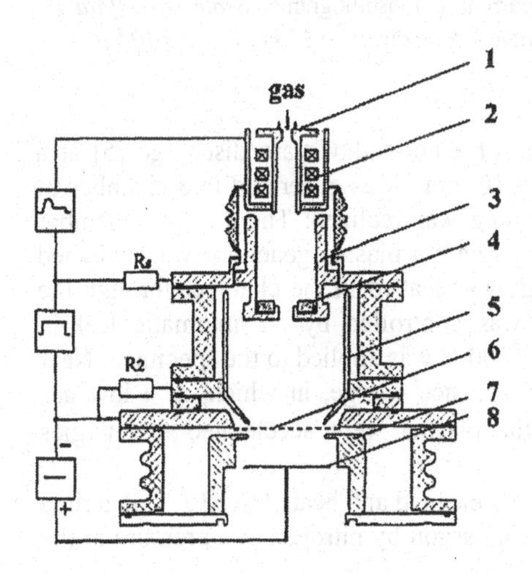

Fig. 8. Pulse plasma generator based on initiated arc by glow discharge.

The parameters of the discharge plasma were measured by a small cylindrical probe and its emissive properties by collector *8* placed at a distance of 5 mm from the grid electrode on application of an extracting voltage.

For the main discharge current equal to 250 A and the grid electrode diameters $d_1 = 5$ cm and $d_2 = 1$ cm, using a small cylindrical probe, we measured the plasma parameters near the grid. The probe measurements have shown that as the anode area is decreased 20 times, the plasma density n increases by an order of magnitude (from $5 \cdot 10^{11}$ to 10^{13} cm^{-3}), the electron temperature T_e decreases from 15 to 10 eV, and the anode fall potential changes from negative (–8 V) to positive (+1 V). On application of an extraction voltage ($U_{extr} = 10$–15 kV) between the grid electrode and the collector, the emission current density reached 10 A/cm^2 in the first case and 100 A/cm^2 in the second one.

The discharge current and emission current oscillograms taken at a pressure $p = 1.5 \cdot 10^{-2}$ Pa are shown in Fig. 9. The fact that the emission current is over the discharge current during the pulse can be due to the gas ionization in the discharge gap and to the ion flow toward the cathode. The maximum current produced in the discharge system is 600 A. This type of discharge is used in low-energy electron sources. The beam current density produced at an accelerating voltage of 20 kV was 120 A/cm^2.

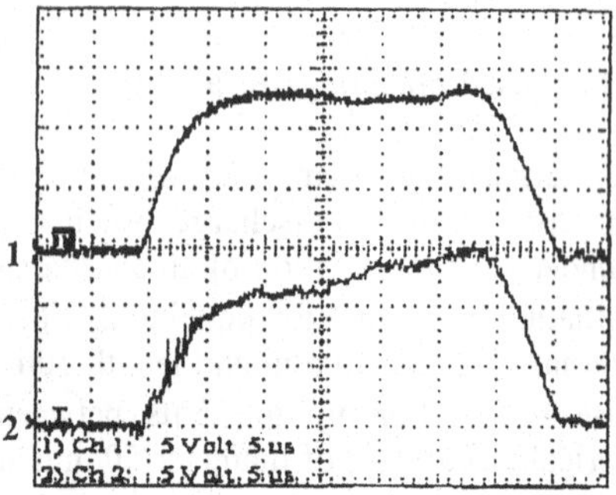

Fig. 9. Discharge current (*1*) and emission current (*2*) oscillograms. *Scale – 100A/div., time – 5 μs/div., V = 17 kV, P = 1.5·10^{-2} Pa.*

5. Applications of plasma generators

Nitriding in arc discharge plasmas.

A structural steel was nitrided in the plasma of a hot-cathode arc discharge [5] at a pressure $p = 1$ Pa. The plasma of density $n = 10^{10}$cm^{-3} was generated in a chamber of volume 0.5 m^3. The following mode of nitriding was realized: The vacuum chamber was evacuated to a pressure $p = 10^{-2}$–10^{-3} Pa where a plasma generator was switched on. The argon gas ionized in an arc discharge was leaked in the chamber through the plasma generator. The argon consumption was controlled by an automatic leak-in system. A negative bias voltage (U_b) of up to 1000 V was applied to the specimen. Near the specimen surface there emerged a layer of space charge, in which, at a low gas pressure ($p \cong 10^{-1}$Pa), ions coming from the plasma were accelerated to energies corresponding to the applied voltage.

When bombarded by Ar ions, the surface was cleaned and heated to 530 °C in a time $t = 1$ hr. Nitriding was performed upon replacing argon by nitrogen at a pressure in the vacuum chamber $p = 3 \cdot 10^{-1}$ Pa. No hydrogen leak-in therewith took place. The test specimens (4140 steel, 0.4 C; 1.0 % Cr) of diameter 20 mm and height 10 mm were annealed to have ferrite-pearlite structure. Before being placed in the working chamber, they were mechanically treated, including polishing, and cleaned of organic contaminants in an ultrasonic tank.

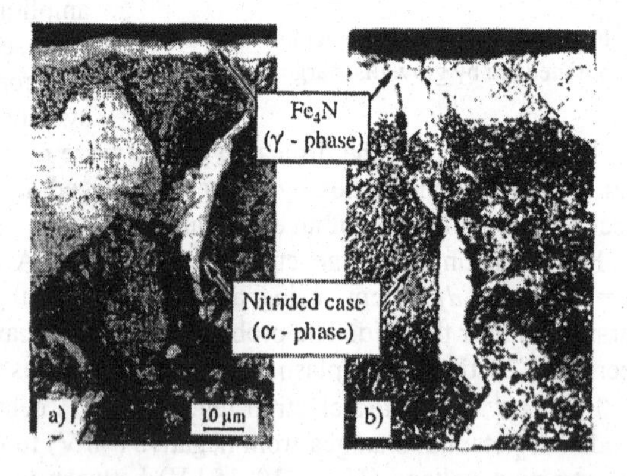

Fig. 10. Microstructure of the 4140 Steel layer nitrided at T = 520 °C for τ = 5 h: in the glow discharge plasma, p = 300 Pa (*a*); in the arc discharge plasma, p = 1 Pa (*b*).

Figure 10 shows the structure of 4140 steel and Fig. 11 shows the variation in microhardness with depth upon nitriding in the arc discharge plasma during 5 hr at a

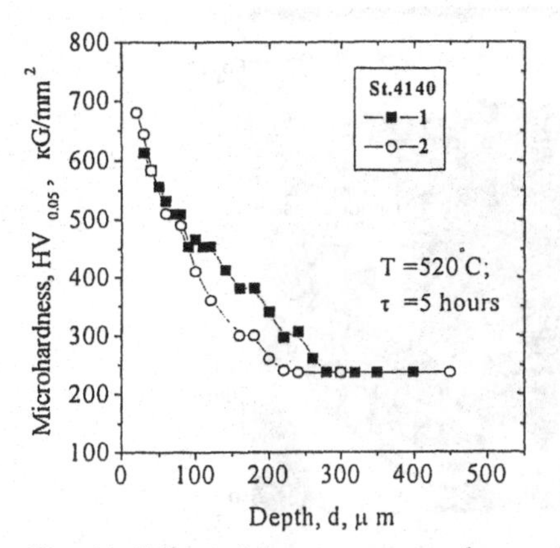

Fig. 11. Effect of the treatment modes on the distribution of the 4140 steel hardness of nitrided layers: upon nitriding in a glow discharge (*1*) and in an arc discharge (*2*).

discharge current $I_d = 50$ A and a bias voltage $U_b = -600$ V. It can be seen that a continuous white layer of the γ'-phase (Fe_4N) is formed at the surface. The width of the layer is 9–12 μm and its hardness is $HV = 7.5$–8 GPa. Under this layer, there is an α-phase (solid solution of N_2 in ferrite) layer whose thickness cannot be evaluated on the microstructural level. This phase consists of light grains mixed with dark pearlite ones. The microhardness of the α-phase varies from 6.5 to 2.5 GPa (pith hardness) within 300 μm.

The thickness of the nitrided layer increases unevenly during the nitriding process. Initially, the γ' and α phases are forming with high speed and then process moderates.

Our results are in agreement with the results reported in [10]. In the experiment described in [10] the nitriding process was performed without hydrogen at a pressure of 10^{-1} Pa and a bias voltage of –400 V. However, as established by our investigations, the nitriding process goes slower if the bias voltage is decreased (Table 3) and ceases if the bias voltage is less than –100 V.

Effect of the bias voltage on the thickness of nitrided layers. **Table 3.**

No.	U_b, V	Thickness of Fe_4N-layer, μm	Thickness of diffusion layer, μm
1.	0	0	0
2.	-80	0	0
3.	-320	7 - 9	65
4.	-600	7 - 9	95
5.	-1000	7 - 9	85

In nitriding, hydrogen is generally introduced in the vacuum chamber to prevent the formation of an oxygen film on the specimen that interrupts the nitriding process [11]. In an arc discharge, the ions accelerated to an energy corresponding to the bias voltage in the space charge layer formed nearly the specimen surface bombard the surface and destroy this film if the ion flow is more intense than the oxygen flow. The diffusion zone is absent and the nitrided layer is invariable if nitriding is performed for a steel which contains no chromium.

Plasma-assisted deposition of coatings. After nitriding in the plasma of an arc discharge, a TiN coating is deposited on the surface of specimens of type 4140 steel. As a result, a modified layer is formed on the surface, which consists of three sequential

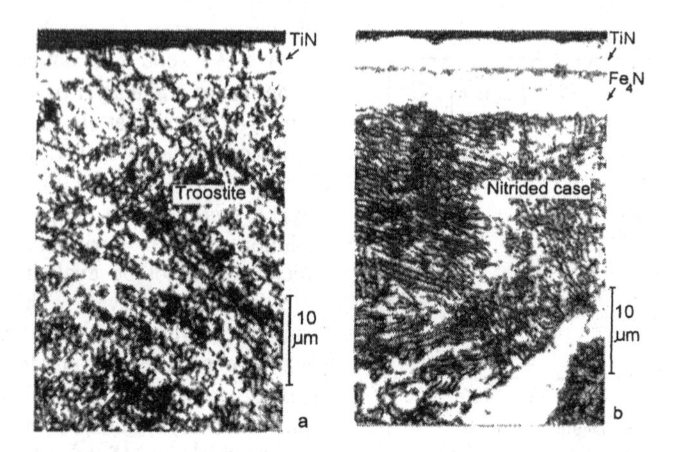

Fig. 12. Structure of the modified layer in 4140 Steel after coating on the water hardening steel (*a*) and complex plasma-assisted ion treatment (*b*).

adhesively bound zones (Fig. 12,*b*). The nitrous ferrite region (α-phase) has a hardness decreasing gradually from 6.0 GPa at the surface to 2.0 GPa in the bulk over a length of 100-200 μm. The above-located ferric nitride (Fe_4N) region of thickness 6 μm has a hardness of 7.5-8.0 GPa. This multilayer composition concludes with a superhard (20 GPa) TiN coating of thickness 4 μm.

Tribotechnical tests have shown that the durability is strongly affected by the hardness and thickness of the coating and by the phase composition of the sublayer. The thickness of the Fe_4N layer on two specimens was 6 μm.

For the identical conditions under which tribotechnical tests were performed for the specimens, it turned out that the protective coating being a three-layer composition (Fig. 12,*b*) shows a wear resistance four times higher than that of the two-layer coating (Fig. 12,*a*).

A correlation between the enhancement of the durability of specimens and the presence of γ'-phase in the diffusion nitrided layer formed in the plasma of an arc discharge has been found. In the absence of such a layer, the durability of the TiN coating on a specimen having only an internal nitriding zone (α-phase) was noticeably lower than the durability of the coating on a specimen containing a thin external Fe_4N zone ensuring enhanced adhesion due to the strong crystal-chemical bindings with the TiN coating and an extended internal nitriding zone. The thermostable layer formed between the structural material and the wear-resistive coating adds to the safety factor for plastic flow and to the rigidity of the structural matrix. This moderates the tendency of the working surface of machine parts to elastic deflection and loss of form stability as a result of plastic deformation under the action of thermomechanical loads. Such loads appear in the contact zone of friction pairs at elevated contact stresses and on cyclic loading of the working surface of the part. If the surface of an article on which a wear-resistive coating has been deposited has an insufficient rigidity, the thin TiN film, notwithstanding its high hardness, is destroyed when coming in contact with the counterbody in the process of friction under elevated specific loads. The extended

transitory layer with a gradually increasing microhardness present in the multilayer coating eliminates the sharp interface between the soft matrix of the substrate material and the superhard working coating, thereby damping the rigidity gradient between these dissimilar materials.
The nitriding of 4140 steel is performed in plasma of arc discharge with hot cathode at discharge current of 50 A and burning voltage of 40 V and pressure of $3 \cdot 10^{-1}$ Pa. The nitric plasma of density 10^{10} cm^{-3} is created in the vacuum chamber of volume 0.5 m^3
The nitriding process is performed as follows: the vacuum chamber is pumped up to pressure $1 \cdot 10^{-3}$ Pa and Ar-gas is leake- in to chamber and the negative bias of – 600 V is applied to specimen at switched on plasma generator.
The thickness of nitriding leer increases unevenly at nitriding process. At the beginning, the and y- phases are formation with high speed and then process moderates. The ours results are agreement with results [10] in which was shown that the nitriding process is performed without introduce of hydrogen at pressure of 10^{-1} Pa and bias voltage of – 400 V. However, as it was established by ours investigations the nitriding process is slow if the bias voltage decreases (table 2) and one ceases if bias voltage is least – 100 V. The diffusion zone is absent and the nitric leers are invariable at nitriding of steel that do not contain of Cr.
Usually at nitriding the hydroding is introduced in vacuum chamber to prevent formation of oxygen film on the specimen that interrupts the nitriding process [11]. In arc discharge the ions bombarding surface spatter this film if ion flow is over flow of oxygen.
Deposition of corrosion-resistive films. The generator of plasma with cold hollow cathode was used for deposition of a corrosion resistance films in the argon – methane plasma at argon-methane relation equal of 1:1. As shown preliminary investigation the transparent amorphous *a*- C:H film with high adhesion and thickness of 0.5 um was formed at the discharge current of 30-40 A and treatment in during of 30 min. The microhardness of film is of 6 GPa on specimen surface and speed of forming is of 3-4 Å/s The increasing of the treatment time leads to decrease of the film forming speed. The film-is not formed if the treatment exceeds some time, that depends on mass of silver specimen.

References

[1] S.P. Bugaev, Yu.E. Kreindel, and P.M. Schanin, Electron Beams of Large Cross Section [in Russian], Energoatomizdat, Moscow, 112 (1984).

[2] N.N. Koval, E.M. Oks, P.M. Schanin, *et al.*, Nucl. Instrum. Meth., **321**, 417-428 (1992).

[3] S.W.A. Gielkens, P.J.M. Peters, W.J. Witteman, et al., Rev. Sci. Instrum., **67**, 2449-2452 (1996).

[4] V.N. Devyatkov, N.N. Koval, and P.M. Schanin, Zh. Tekh. Fiz., **68**, 44-48 (1998).

[5] D.P. Borisov, N.N. Koval, and P.M. Schanin, Izv. Vyssh. Ucheb. Zaved., Fiz., 115-120 (1994).

[6] D.P. Borisov, N.N. Koval, P.M. Schanin, RF Patent No. 2116707, Byul. Izobr., No. 21 (1998).

[7] D.P. Borisov, I.M. Goncharenko, N.N. Koval, and P.M. Schanin, IEEE Trans. Plasma Sci., **26**, 1680-1684 (1998).

[8] P.M. Schanin, N.N. Koval, I.M. Goncharenko, and S.V. Grigoriev, Fiz. Khim. Obr. Mater., **3**, 16-19 (2001).
[9] A.V. Belyi, V.A. Kukarenko, I.Yu. Tarasevich, *et al.*, *Ibid.*, **4**, 11-17 (2000).
[10] Sanchette F., Damonde E., Burvron M., and *et. al.*, Surf. Coat. and Technol., **94 - 95**, 261 – 267 (1997).
[11] S. Parancandola, O. Kruse, and W. Moller, J. Appl. Phys. Lett., **75**, 1851 (1999).

ELECTRON BEAM DEPOSITION OF HIGH TEMPERATURE SUPERCONDUCTING THIN FILMS

G.Mladenov, K.Vutova, G.Djanovski, E.Koleva, V.Vassileva, D. Mollov
Institute of electronics at Bulgarian Academy of Sciences, Sofia 1784, 72 Tzarigradsko shose, Bulgaria

Abstract. Some results of characterization of prepared high temperature superconducting HTS (namely $YBa_2Cu_3O_7$) thin films by electron beam physical vacuum deposition (EB PVD) are given. An approach for the use of high rate EB PVD of the (HTS) depositing layers is discussed. A need of clarification of: (i) the role of the used electron beam energy distribution at the melted and vaporized material surface on the molten metal stirring, (ii) the effect of the temporal (not continuos) thermal contact between evaporated material and crucible as well as (iii) the consequences of the processes of the beam interaction with the vapors on the obtained evaporating rate is discussed.

1. Introduction

High temperature superconducting (HTS) thin films (of type of $YBa_2Cu_3O_7$) have various prospective uses. Example of an important microwave electronic application is narrow bandwidth channel filters in the mobile cellular telephone base stations and in wireless communication in general. Another prospective utilization of these films is the superconducting quantum interference device (SQUID). Magnetometers, applied as example in medical diagnostics or for nondestructive detection of metal defects in the civil and the aerospace industry are developed on its base. There is also worldwide interest in design and applications of various HTS sensors for measurements of electromagnetic radiation and flows of accelerated particles (atoms, molecules and clusters).

In addition to the development of system concept and device design of such HTS products there is an urgent need to develop the technology for achieving higher quality at a less expensive and larger scale production route. Now the manufactured film area is usually 0.25-2.5cm^2 / deposition / sample but it is needed 1-10 thousand larger area. For the needs of image processing (for example in the IR region) a big improvement of the multi-pixel HTS matrix mastering is desirable. This means that there is a hard need of improvement of the used lithography technology.

In this paper some general problems of HTS thin film technology are discussed on the example of the electron beam physical vacuum deposition (EB PVD) of the HTS

E. Oks and I. Brown (eds.),
Emerging Applications of Vacuum-Arc-Produced Plasma, Ion and Electron Beams, 163–171.

films (of type $YBa_2Cu_3O_7$) deposited on substrates of MgO (MO), $SrTiO_3$ (STO) or $LaAlO_3$ (LAO) with matched crystal structure and thermal coefficient of expansion.

2. Experimental investigation of the $YBa_2Cu_3O_{7-\delta}$ thin films

As it is known [1] (see fig.1) the $YBa_2Cu_3O_{7-\delta}$ has an orthorhombic unit cell (comprised of three perovskite-like sub-cells that are not quite cubic). In a unit of that cell the central atoms are barium on the end sub-cells and yttrium on the central sub-cell, while copper occupies all of the corner sites. Oxygen deficiency is notable for this superconducting ceramic. The precise number of oxygen sites, which are unoccupied, is crucial to the superconducting behavior of the compound. Changing the oxygen deficiency level δ, one can completely change the electrical properties of the material ranging from a superconductor ($\delta=0$) to an anti-ferromagnetic insulator ($\delta=1$).

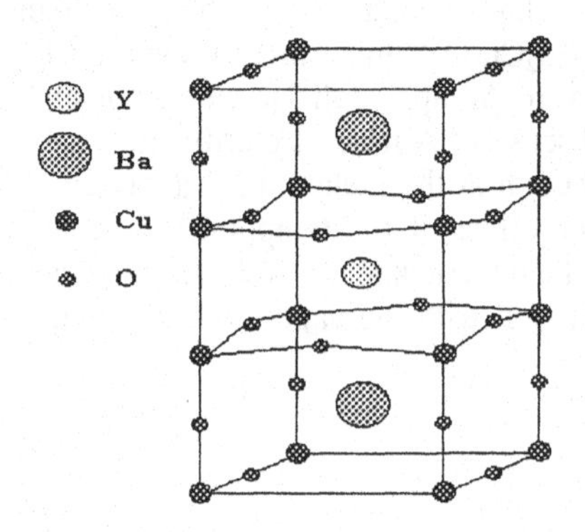

Figure 1. The $YBa_2Cu_3O_{7-\delta}$ unit cell.

Decreasing the oxygen content, the$YBa_2Cu_3O_{7-\delta}$ crystal structure simultaneously undergoes a transition from orthorhombic ($a<b$) to tetragonal ($a=b$). For $\delta<0.6$ the $YBa_2Cu_3O_{7-\delta}$ exhibits a metallic behavior with critical temperature T_c reaching 91 K for $\delta=0.03$. For $0.6<\delta<0.8$ the material shows semiconductor like characteristics (and found applications at room temperatures [2]) and for values of $\delta>0.8$ the compound behaves as insulator at low temperatures.

The preparation of investigated in this paper HTS films of type of $YBa_2Cu_3O_{7-\delta}$ were done by a prototype equipment consisting of a 3 kW electron gun (and connected high voltage electrical source and suitable electronic control systems) together with three water cooled crucible system or with an especially designed novel hot multi-pot evaporation set with 8 copper and Graphite pots (see patents [3,4]) as well as an SiC substrate holder heater. The samples were annealed for 240 min in the deposition chamber at 700°C (heating time 20min; cooling time 180min) at oxygen pressure 200 Torr.

The following control operations were done during the deposition of $YBa_2Cu_3O_{7-\delta}$ layers:

- monitoring the film component composition using pre-measured rates of deposition of every individual component at a chosen electron beam power density on its evaporation source. In these preliminary experiments the method of the oscillating quartz was utilized as a measurement method for the estimation of the deposited component masses;
- monitoring the film component composition using a optical system for observation of light emission from the exited vapor atoms during the electron beam evaporation (Cu - 324.75 nm; Y - 437.49 nm and Ba - 455.4 nm).
- measurement of plasma parameters at low EB powers [5,6,7].

After deposition of the produced films their coating compositions were investigated by:

- X-ray diffraction (XRD),
- SEM analysis, including energy dispersion X-ray microprobe analysis (EDX),
- spark spectral photometer (destructive analysis).

The deposition of $YBa_2Cu_3O_{7-\delta}$ films is a multi-factor process. The most significant factors are: (i) atomic composition control by EB parameters and thus produced evaporating rates; (ii) substrate temperature; (iii) crystal parameter matching between the substrate (or buffer layer) and the grooving HTS film; (iv) post-deposition baking cycle for adjustment of the oxygen component in tight tolerances.

Examples of XRD spectra are shown in fig.2. They are from $YBa_2Cu_3O_{7-\delta}$ films

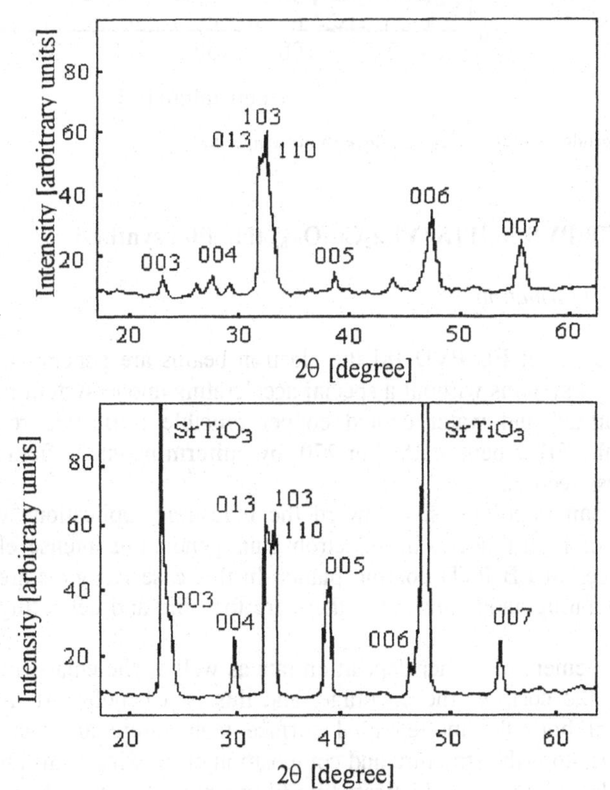

Figure 2. X-ray diagram of $YBa_2Cu_3O_{7-\delta}$ film deposited on Al_2O_3.

deposited on Al_2O_3 - fig.2(a), and $SrTiO_3$ - fig.2(b). Figure 3 shows an example of the dependence of film electrical sheet resistance R (measured in Ω of a square) on the temperature T (measured in K). The film thickness is 100nm. The resistivity is $\rho \approx 1$ μΩ.m at the sheet resistance 10 Ω/□ observed in the region of the threshold from normal to superconductive state. Four points measurement method at dc current 50μA and at zero value of the applied magnetic field shows a sharp resistance transition at temperature T_c equal to 89K. The film dc resistance remains zero below the transition critical temperature and at higher temperatures exhibits a normal resistance level.

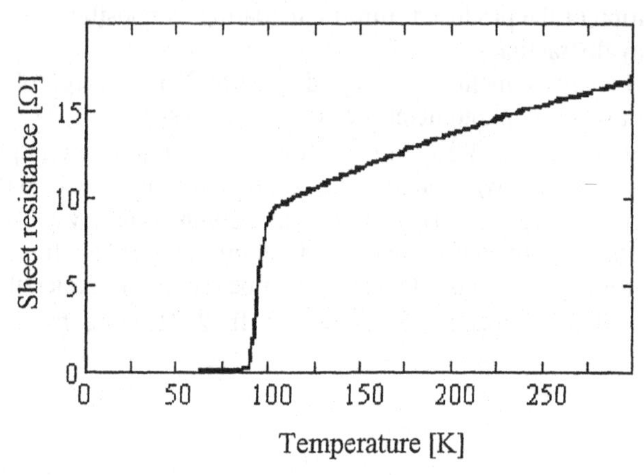

Figure 3. Dependence of the sheet resistance on the temperature.

3. Use of EB PVD in HTS ($YBa_2Cu_3O_{7-\delta}$) thin film synthesis

3.1. General information

In the most cases of EB PVD [8] the electron beams are generated by simple radial electron optical systems without a special accelerating anode system (since the material to be evaporated and water cooled copper crucible fulfil this role). In these EB evaporators the EB is bent at 180° or 270° by uniform magnetic field and often have a strip-like cross section.

Recently an important step toward the industrial applications was done by the introduction of high power axial electron guns generating intense electron beams as heating sources in EB PVD coating plants. In this case EB guns are completed with special accelerating anode and often there are focusing and deflecting electromagnetic systems.

The achievement of higher deposition rate as well as the enlargement of deposition area reduces the costs of the coverage, and this is a benefit. At the same time the instabilities of both the molten pool surface temperature and the evaporation rate leading to variations the structure and composition of growing films are problems in the case of EB deposition of sophisticated multi-component films. As a consequence the desirable uniformity of the properties of the deposited material can be deteriorated.

A stabilisation of the liquid flow dynamics in the molten pool can be found by: (i) proper choice of beam trajectory on the molten pool surface, (ii) application of pulse regime of evaporation or (iii) using of intermediate metal heater [9,10] as methods of obtaining higher rate of deposition at sufficient quality.

Beside the high rate deposition effects the created film structure grows worse. The methods permitting the improvement of the layer properties remarkably are the use of ion-assisted deposition or plasma activated growing of coating [10]. This method has to be tested, but has not been applied yet. In HTS film mastering a cluster beam bombardment is used for smoothing the surface [11] followed by annealing in oxygen ambient to regrow the damaged surface.

The concurrent processes to the EB evaporation and deposition of HTS films are the magnetron sputtering, the laser ablation and the plasma spray deposition. All these technologies use sintered HTS samples as a raw material and process aim during the deposition is to transfer this previously prepared material composition at some improvement or without destroying the crystal structure in the grooving layer.

From the other side, the EB PVD of HTS films on the substrates uses consequent or simultaneous vaporization of the film metallic components, synthesizing the desired chemical composition and the crystal (or granular) structure of the film during its deposition itself. The oxygen content in the film is adjusted precisely by an additional baking of the film in oxygen atmosphere. This approach of film deposition is many times less expensive than in the case of all other mentioned methods. This is so because the other methods utilize complicated (and expensive) pretreated raw materials.

The use of an extracted from the vacuum arc plasma electron beam as a heating source in the PVD equipment, designed for the complex oxide deposition, seems to be of benefit. Typical heat power and its distribution, beam control ability to evaporate materials with different evaporation rates and mainly sustainable work of such an evaporation and deposition system in oxygen environment (with relatively high oxygen pressure - about 10^{-1} – 1 Pa) are important advantages of these devices.

EB evaporation and deposition processes permits also the fabrication of other sophisticated multi-component layer systems and new materials (fero-electrics, fero-magnetics and electro-optical or bio-compatible ceramics). Optimisation of the coating parameters and a derivation of the benefits from the capabilities of EB technology relies on a deep knowledge of the processing phenomena.

3.2 Discussion on evaporation rate at EB PVD

The relation of the evaporation rate M_v on temperature of the molten pool surface T in any point of that surface is given by [12]:

$$M_v = C_1 (M/T)^{1/2} . \exp(-C_2/T) , \quad (1)$$

where M is the molecular weight of the evaporant and C_1 and C_2 are constants .

The relation of temperature T on beam power P has to be discussed. If one tries to measure the surface temperature of the molten pool using a color optical pyrometer the obtained data are for a mean surface temperature, averaged for given period and area, instead the measures of the local momentum over-temperature of the metal surface in the beam spot. So, evaluations of evaporating rate using that experimental temperature are too rough. Another rough estimation can be based on theoretical results in [13], obtained for the heat transfer through liquid turbulent flow and thermal conductivity with superimposed thermal capillary effect. In agreement with these results the temperature rise is proportional to the 2/3 power of the beam power density. Some other values can be found using computer simulation [14-18]. They are also in rough figures due to difficulties in exact prediction of both: (i) turbulence flow of molten metal (controlled by thermo-capillary and convection forces) and (ii) of the heat contact conditions in the interface material/crucible wall.

At increase of the EB power of an evaporator used for PVD of thin films one can expect a monotone increase of the rate of evaporation and in this way the deposition rate of the material. An example of experimentally observed evaporation rates at different EB powers in case of evaporation of copper at EB powers going to 30 kW and a diameter of crucible 6 cm; as a steady electron beam situated on central part (ϕ20mm) of the molten pool surface is shown in fig.4. A saturation of the observed evaporation rate at increase of EB power occurs. It is shown that a maximal EB power in which evaporation efficiency is optimal exists. The value of this optimal EB power is

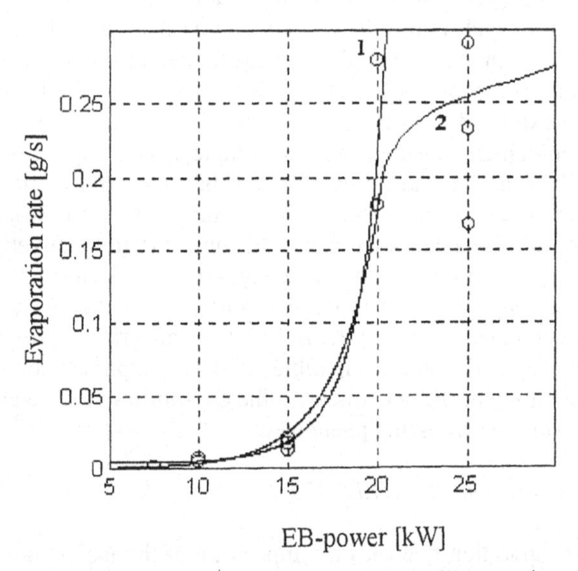

Figure 4. Evaporation rate v.s. EB-power. (1) – theoretical curve; (2) – experimental curve using average values; (o) – momentum values of the evaporation rate measured by an optical emission method.

correlating with the material parameters, the energy distribution of the beam, its movement trajectory and velocity, with the diameter of the melted block (and the crucible inner diameter), with the length of that block and probably with the other features of the concrete EB evaporating system. The near to saturated values of evaporation rates at higher EB powers can be explained: (i) by non-linear heat conductivity of the system consisting of molten pool and crucible walls or (ii) by an intensification of the electron beam - plasma interactions in the vapor - plasma cloud above the beam spot, as well as (iii) with the generation of a concavity due to the impulses of the ejected evaporated atoms and to the vapor pressure, acting as additive reason for (i).

The choice of the movement trajectory and speed, as well as the space and the energy distribution of the beam is an important way to control the evaporation rate at given EB power. To change the over-temperature (temperature elevation) of the molten metal surface in the region of the EB spot one can change the rate of the beam spot motion and its trajectory. The temperature distributions on the rest part of the molten pool surfaces and in the depth of the liquid metal are also important factors. The molten metal temperature distribution is a base factor controlling the evaporation rate and the liquid flow and heat transport in the molten material. The geometry parameters of the molten pool and of the beam spot are also important.

As an attempt to generalise the effect of different beam trajectories and energy distributions, the main conclusions can be listed as follows:

(i) The scanning frequency has strong effect on the surface temperature. It changes shape of the cross section of the molten pool and no practical effect on the depth of the molten pool (usually measured along the crucible axis).

(ii) When increasing the movement beam speed at rise of beam deflection frequency a decrease of the temperature elevation of this hot spot is observed. For frequencies more than 10 Hz (to 10 kHz) the temperature surface distribution of the molten pool is practically coinciding with the quasi-stationary regime. At lower frequency values a high evaporation rate is observed. This means that this region of beam deflection less then 10 Hz is suitable for EB PVD and the frequency values more then 50 Hz are useful for EB melting. In the case of low frequencies (EB PVD) the trajectory of the beam movement controls the liquid metal flows and heat losses to the crucible. Ring movement of the beam is better for evaporation because the cold liquid metal is keeping in the peripheral zone of the molten pool. Closed and simpler trajectories permit the liquid flow in the molten pool to be more laminar, then turbulent. At these conditions surface state waves are created probably.

(iii) The liquid surface close to beam impact is deformed due to the downward thrust on the surface from the static pressure of the vapour and the dynamic pressure of its momentum, balanced by capillary and hydrostatic pressures of the liquid [19]. The surface profile affects the temperature distribution in the liquid pool. The surface of the pool goes through time fluctuations (small waves) due to the temperature time variations.

(iv) The main part of evaporated material is originated by EB spot (at steady beam or at low frequencies of its scanning movement on the molten pool surface).

(v) Conclusion for a lack of a steady state thermal convection of liquid metal is extracted from computer simulations [16]. Some times the movement of molten flows is approximated as laminar but normally they have a semi-chaotic behaviour. Other simulations [14,15] show a non-steady thermal contact conditions at interface between inner wall of water cooled crucible and side wall of the cast ingot.

(vi) Some beam vapor interaction processes are also reason for high frequency oscillations of the beam power in the heating spot (scattering and gas focusing of the beam, oscillation of vapor plasma parameters etc.).

4. Conclusions

HTS films were successively produced by EB PVD on various samples with suitable crystal structure. Results are obtained at EB powers up to 3 kW. The need for exploring high rate EB PVD is discussed. Some reasons concerning observed saturation of the evaporation rate at high beam powers are also mentioned. Stabilisation of the surface temperature of the molten metal pool and/or improvements of the deposition rate measurement are possible approaches. Unfortunately the final decision for better way to overlay the difficulties due to non-steady evaporation rate was not definitely done.

Acknowledgements

The research was funded through Science for Peace NATO program (SfP 973718 project) and the National Council "Scientific Investigations" at the Ministry of Education and Science of Republic of Bulgaria under contract No. MM-1004.

References:

1. Doss, J.D. Engineer's *Guide to High-Temperature Superconductivity*, John Wiley & Sons, NY, (1989).
2. Z.C.Butler, P.C.Shan, D.P.Butler, A.Jahanzeb, C.M.Travers, W.Kula, R.Sobolewski, *Solid-State Electron.* **41**, 895 (1997).
3. V.Laska, A.Grishanov, V.Luchinin, V.Koulkov, A.Ilin, G.Mladenov, N.Dimitrov, T.Djacov, Method for fabrication of HTS thin films, *USSR patent certificate 4787061, Registration No 25013189 from 30.01.1990.*
4. V.Laska, A.Grishanov, V.Luchinin, V.Koulkov, A.Ilin, G.Mladenov, N.Dimitrov, T.Djacov, Method and equipment for deposition of the HTS coatings, *Certificate for Bulgarian invention No 49914, Registration No 91033 / 26.01. 1990.*
5. Dyakov T., Bielavsky M., Kardjiev M., Djakov B., Mladenov G., Electric probe studies of the ionised metal vapour accompanying electron beam welding, in the *Proc. of the First Intern. Conf. on Electron Beam Technologies EBT-85*, Publ. House of BAS, Sofia, pp.199-204 (1985).
6. G.Mladenov, S.Sabchevski, *Vacuum* **62**, 113 (2001).
7. H.Kajioka, *Vacuum* **48**, 893 (1997).
8. R.T.Bunchah, *Z.fia-Metatlkunde*, **75**, N.I 1, 840 (1984).
9. G.Mladenov, N.Atanasov, *Certificate for Bulgarian invention No 23870 (1976).*
10. V.Vassileva, E.Georgieva, E.Koleva, Acceleration of electron beam evaporation, in the Proc. of National Conf. "Electronica'98", Publ. UEEC, pp.70-72 (1998).
11. W.Chu, Y.Li, J.Liu, J.Wu, S.Tidrow, N.Toyoda, J.Matsuo and I.Yamada, *Appl.Phys.Lett*, **72** (2), 246 (1998).
12. C.Melde, M.Kammer, A.von Ardene, M.Neumann, The super deflection system – a tool to reduce the evaporation losses in EB melting processes, in R.Bakish (ed.), *Proc. of the Conf. Electron beam melting and refining State of the Art 1993*, Bakish Materials Corporation Publ., Englewood NJ, pp.69-79 (1993).
13. J.A.Pumir, L.Blumenfeld, *Phys.Rev.*, E. **54**, R4528 (1996).
14. K.Vutova, V.Vassileva, G.Mladenov, *Vacuum* **48**, 143 (1997).
15. E.Koleva, K.Vutova, G.Mladenov, *Vacuum* **62**, 197 (2001).
16. H.S.Kheshgi, P.M.Gresho, Analysis of electron-beam vaporization of refractory metals, in R.Bakish (ed.), *Proc. of the Conf. Electron beam melting and refining State of the Art 1986*, Bakish Materials Corporation Publ., Englewood NJ, pp.68-79 (1986).

17. K.W.Westerberg, M.A.McClelland, B.A.Finlayson, Numerical simulation of material and energy flow in an e-beam melt furnace, in R.Bakish (ed.), *Proc. of the Conf. Electron beam melting and refining State of the Art 1993*, Bakish Materials Corporation Publ., Englewood NJ, pp.153-165 (1993).
18. J-P.Bellot, E.Floris, A.Jardy, D.Ablitzer, Numerical simulation of the E.B.C.H.R. process, in R.Bakish (ed.), *Proc. of the Conf. Electron beam melting and refining State of the Art 1993*, Bakish Materials Corporation Publ., Englewood NJ, pp.139-152 (1993).
19. A.Mitchell, H.Nakamura, D.W.Tripp, The "hot-spot" problem in EB melting of alloys, in R.Bakish (ed.), *Proc. of the Conf. Electron beam melting and refining State of the Art 1987*, Bakish Materials Corporation Publ., Englewood NJ, pp.23-32 (1987).

DEPOSITION OF NANOSCALE MULTILAYERED STRUCTURES USING FILTERED CATHODIC VACUUM ARC PLASMA BEAMS

M.M.M. Bilek[1,2], D.R. McKenzie[1], T.W.H. Oates[1], J. Pigott[1], P. Denniss[1] and J. Vlcek[2]

[1] *Applied and Plasma Physics Group, School of Physics, University of Sydney, NSW, 2006, Australia.*

[2] *Department of Physics, Faculty of Applied Sciences, University of West Bohemia, Plzen, Czech Republic.*

Abstract. Nanoscaled multilayered thin film structures have recently been identified as good candidates for applications where super-tough and super-hard materials are required. The properties of the structures diverge considerably from those of the bulk constituents and thus are likely to depend on the interfaces present. Interfaces in such structures can be smooth, graded or rough. To determine to what extent the interface quality determines the properties it is necessary to have control of the degree of mixing at the interface during deposition. The fully ionized cathodic arc plasma has a relatively narrow natural ion energy distribution as compared with other deposition processes such as sputtering. This affords good control of the impact energy at the growth surface using substrate bias, enabling control of the degree of interface mixing. This paper discusses a dual source pulsed filtered cathodic arc plasma deposition system designed to deposit a variety of multilayered thin film materials with various interface morphologies. We discuss also methods of producing multilayered materials in single source continuous arc systems.

1. Introduction

Currently available coatings for protective and mechanical applications that consist of a single homogeneous material suffer from deficiencies in adhesion, fracture toughness and friction coefficient. No single material is optimum in all three aspects. The range of available coatings can be greatly extended by the use of composites. Nanoscale multilayer films have been shown to provide enhancements in indentation hardness and fracture toughness beyond those of the constituent materials [1,2]. Because of the large volume fraction of the interfacial regions in these structures it is likely that these regions will provide the key to understanding these unusual properties. The structure of the interfacial regions will depend primarily on the energies of incident species on the growth surface and the temperature of that surface as these factors govern the level of

E. Oks and I. Brown (eds.),
Emerging Applications of Vacuum-Arc-Produced Plasma, Ion and Electron Beams, 173–186.

mixing and diffusion. The majority of nanoscale multilayered structures so far reported have been produced using techniques such as sputtering in which only a small fraction of the condensing species are ions and the energy range of incident species is relatively broad, as given for example by the Thompson distribution of sputtered atoms [3]. The cathodic arc in comparison produces a narrow energy distribution consisting almost entirely of ions. This opens up the possibility of fine control of the structure of the interfacial regions by using electrical bias to tune the energies of incident species.

The cathodic arc is a source of highly ionised plasma of metals and other solid elements, such as graphite, with sufficient electrical conductivity to sustain an electric arc. It is now widely used for many applications in materials synthesis and processing. An electric trigger is used to start an arc burning on the surface of a solid, electrically conducting cathode. The plasma is produced in small micrometer sized regions called cathode spots where high currents ablate the solid surface and ionise material into a plasma plume. Because of its high level of ionisation the plasma can be manipulated by electric and magnetic fields to change its energy and direction. A major disadvantage of the cathodic arc is that solid and molten debris known as *macroparticles*, which are a by-product of the violent processes occurring in the cathode spots, contaminate the plasma. Since macroparticles have similar or greater dimensions than the desired layer thicknesses in nanoscale multilayered structures, they would seriously compromise the structure and must be eliminated from the plasma. A curved magnetic filter is often used to remove the macroparticles by deviating the plasma through a curved magnetized duct that does not allow a line of sight transit between the cathode and substrate.

Two forms of the cathodic arc have been developed for materials processing applications. They are known as *continuous* or *dc* and *pulsed cathodic arcs*. They involve different methods of triggering and different types of power supplies. The *pulsed arc* uses a high voltage surface flash over across an insulator to the cathode to initiate the discharge and is sustained by current drawn from a capacitor bank, which is recharged between pulses. The continuous (dc) arc is usually ignited by mechanical contact between two electrodes that are then pulled apart again. It is sustained by a dc power supply such as that used for welding. Because of the need to supply current continuously the current is much lower (of order 10s or 100s of Amperes) in the dc arc case.

2. Multilayers

We have used a magnetically filtered dc cathodic arc source to deposit multilayered structures from a single carbon cathode. The plasma immersion ion implantation (PIII) technique combined with vacuum arc deposition (PIII&D) [4] was used to reduce the intrinsic stress and allow the deposition of a thick and adherent film [5]. A layering effect was achieved by interrupting the deposition periodically. This resulted in layers of different microstructure and density. Fig. 1 shows an electron micrograph of the cross-section through a film produced with 30-second bursts of arc current alternated with 30-second rest intervals. The structure appears to relax during the rest periods forming denser layers, which appear dark in the TEM image. The two thicker dark layers correspond to extended rest periods. Compared to similar non-layered coatings this material showed enhanced performance in a pin-on-disk wear test, indicative of enhanced fracture toughness.

Figure 1: Transmission electron microscope (TEM) image showing the cross-section of an 11 µm thick carbon film made up of approximately 200 layers. The layers, which can be seen as alternating light and dark stripes, where produced by alternating 30-second bursts of deposition using the PIII&D technique in a single source dc filtered cathodic vacuum arc with 30-second rest periods. The denser (dark) carbon layers were formed during the rest periods.

This work, however, revealed some limitations of using a single source dc cathodic arc to make multilayered coatings. Firstly, the single source limits the range of composition of constituent layers to the cathode material and its compounds with reactive gases. Secondly, the density of the dc plasma plume fluctuates strongly with the instantaneous arc current. This makes the deposition of precise layer thicknesses at the nanoscale impossible. In this paper we therefore describe the design and characterisation of a pulsed cathodic arc system for the deposition of multilayer structures. As part of this study we examine the operation of the prototype pulsed arc system including the effects on cathode erosion and plasma composition of varying the pulse current profile.

3. Development Of Dual Source Pulsed Cathodic Arc

3.1. SYSTEM DESIGN

In order to deposit multilayer films with a wide variety of constituent materials without breaking vacuum, a system with two sources is required. Our basic system design is shown in Fig.2.

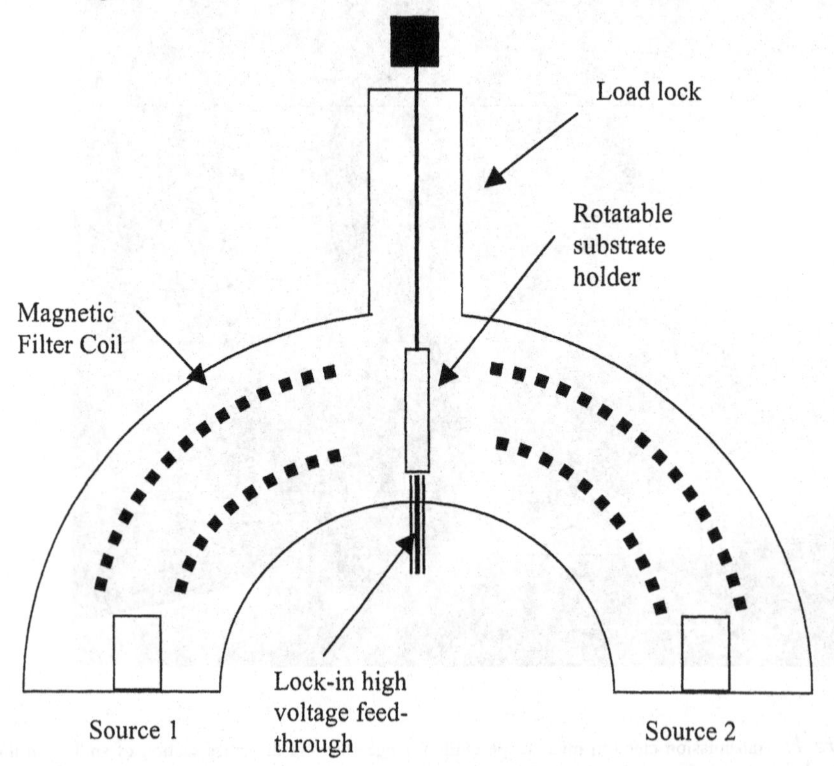

Figure 2: Schematic diagram showing the main features of a dual source filtered cathodic arc system for the production of multilayered thin film materials. Two cathodic arc sources each with a curved magnetic macroparticle filter are placed on opposite sides of a rotatable substrate holder, which can be made to face either of the sources in turn. The lock-in high voltage feed-through allows the use of high-voltage biasing techniques such as PIII during layer deposition.

As we wish to make multilayered films with nanoscale bilayer periods, macroparticles cannot be tolerated and so efficient macroparticle filters are an essential part of the system. A curved magnetic macroparticle filter is placed between each of the sources and the substrate holder. The filters are bent solenoids wound from copper piping. The filters and sources are mounted on curved rails with insulating supports and electrically connected in series with the arc source. The rails allow the entire source-filter structure to be assembled outside the chamber and then simply be slid into place when complete. The rails also allow movement of the source-filter assemblies to almost any position along the segment that they subtend. This allows us to vary the length of the filters without changing the distance between the filter exit and the substrate.

The substrate holder is placed on a mechanical rotatable feed-through so that the substrate can be made to face both of the sources in turn without breaking vacuum. A load lock is used to insert and remove substrates to reduce the turn around time between the coating of specimens. The substrate holder clips into a lock-in high-voltage feed through which can be used to bias the substrate and hence shift the energies of the incoming ions as desired. Good control of impinging ion energy during growth is required to tailor the mixing and/or grading at the layer interfaces. The lock-in feed-through also helps to stabilise the substrate during the deposition of the layers for in-situ measurement of optical properties with an ellipsometer. The ellipsometer ports are mounted in a plane perpendicular to that of the diagram and so are not shown in the figure.

3.2. SOURCE DESIGN

The schematic diagram of Fig. 3 shows the design of the pulsed arc sources. Because of the intrinsically variable deposition rates of continuous arc sources the fine control of layer thickness required for the fabrication of nanolayered structures is difficult to achieve with a dc source. Pulsed current sources however, are capable of producing well-defined, very repeatable arc current amplitudes and pulse lengths. Fine control can thus be achieved by calibrating the fraction of a monolayer deposited during a pulse and then simply counting pulses delivered during the deposition of each layer. An additional benefit of using a pulsed source is the significantly reduced level of macroparticle emission [6].

The design of our pulsed arc sources is essentially based on that presented by Siemroth et al [6] for a high current pulsed vacuum arc. The cathodes are disks with diameter of 5cm and a hole in the centre for the trigger. The trigger is a tungsten wire inserted into an insulating sleeve made of alumina. A high voltage pulse (>1kV for 3μs) applied between the trigger wire and the cathode by the triggering circuit causes a flash over to occur across the insulator, which ignites an arc that burns between the anode and cathode. The capacitors of the pulsed arc power supply then provide the current for the arc.

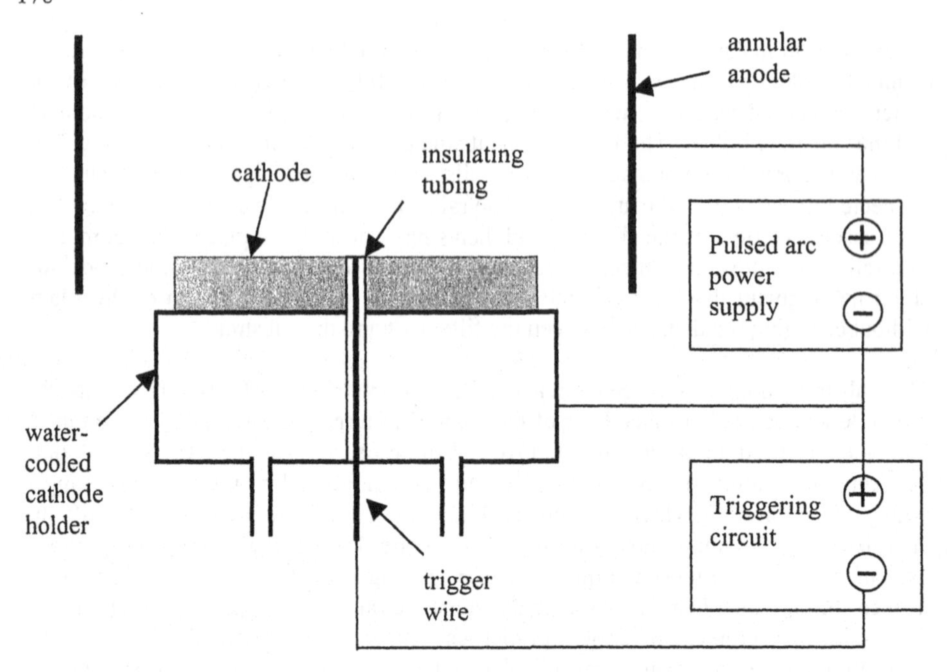

Figure 3: Schematic diagram showing the design of the pulsed arc sources.

3.3. POWER SUPPLY DESIGN

In a cathodic arc the evaporation and ionisation occurs in the hot, localised, high-density plasma clouds close to the cathode surface known as *cathode spots*. The cathode spots move over the cathode surface and generate the plasma which then streams away from the cathode surface as a *plasma plume*. In the early stages of prototyping, our power supply consisted of a simple capacitor bank of between 6 and 12 mF, charged to between 100 and 400V. This provided current pulses with a fast rise-time and a long decaying tail, such as shown in Fig. 4 (a)). The number of cathode spots on the cathode surface increases with arc current and the spots repel one another and travel towards the outside of the cathode during the pulse [7]. A time decreasing current profile therefore results in an excessively high erosion rate near the centre of the cathode where the arc spots are triggered and a very small erosion rate near the circumference of the cathode.

In an attempt to even out the erosion profile across the cathode surface, we designed our next prototype power supply based on an oscillating LC circuit such as that used by Siemroth et al [6], adapted to make it suitable for electrolytic capacitors. Electrolytic capacitors are less expensive for a given capacity than other types, but require a circuit design that will prevent them from being reverse biased. The circuit diagram for our

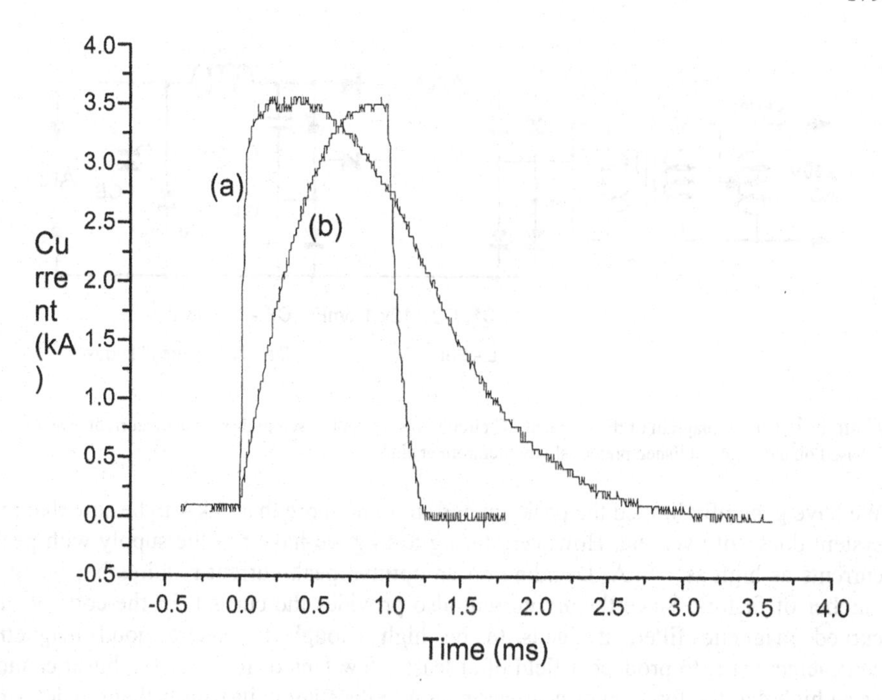

Figure 4: Comparison of current profiles for the two power supplies we tested, (a) the simple 12mF, 400V capacitor bank and (b) the 300V LC pulse circuit, crowbarred at 1ms.

power supply is shown in Fig. 5. The crowbar is used to short out the plasma and cut the pulse current at a predetermined time during the positive cycle (0-180 degrees) of the oscillation. If it is cut prior to 90 degrees into the oscillation cycle, a current pulse that rises over almost the entire pulse length, as shown in Fig. 4 (b), is produced. The crowbarring time can be adjusted to suit the velocity of the cathode spot so that the maximum coverage of the cathode surface is achieved but the spot does not run over the edge of the cathode surface. The crowbar diverts the current from the plasma but keeps the remaining energy in the circuit. When the current drops to zero the crowbar turns off and the "efficiency" diode takes over, and it carries the entire negative cycle of the current. During this time the remaining energy in the circuit is returned to the charged half (C1) of the capacitor bank.

Some of the design issues we are addressing are the trade-offs between peak pulse current, pulse length and repetition rate. Since the aim is to get the highest possible deposition rate for a given power, we would like to maximise the time-averaged ion current and hence the peak arc current and duty cycle.

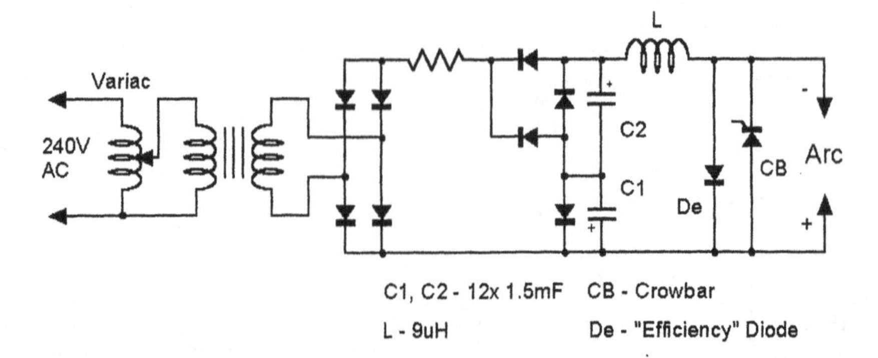

Figure 5: Circuit diagram of the resonant LC circuit power supply used to drive the arc current. The circuit is based on a design published previously by Siemroth et al [6].

We have generally limited the peak ion current to no more than 5 kA to be sure that the system does not overheat. However, during testing we have run the supply with peak currents as high as 10 kA. Deciding on an optimal peak current requires balancing a number of factors. Since the arc current also provides the current for the coils of the curved magnetic filter, it needs to be high enough to ensure good magnetic confinement (i.e. to produce a field of at least a few tens of milli-Tesla), but it cannot be so high that the force on or heat dissipation in the filter coils causes them to deform. High arc currents are advantageous because they generate faster moving arc spots, which produce fewer macroparticles [8], but unfortunately losses in the power supply scale as the square of the peak current. With these factors in mind we designed the power supply to operate with peak currents in the range of 1 to 5 kA.

If the energy of the arriving ions is to be well controlled by the application of a bias voltage to the substrate it is important that the plasma not only be fully ionised but that the ion charge state distribution (CSD) is as narrow as possible and well known. Ions with different charges will experience unequal deceleration by the application of a bias to the substrate and subsequently exhibit different kinetic energies upon arrival at the substrate. Oks et al [9] showed that the CSD is affected by both the application of a strong axial magnetic field at the cathode surface, and also by magnitude of the arc current. For a carbon cathode in the absence of an external magnetic field, currents above 2kA linearly increase the average charge state. Thus, although high current arcs can provide a high deposition rate and reduce the macroparticle content, it is important to consider the effects of the current on the CSD if the ion arrival energy is to be manipulated by an electric bias. This is also true for the application of an axial magnetic field for the purposes of macroparticle filtering and/or cathode spot motion control.

In our system the pulse length is dictated by the radius of the cathode and the velocity of the cathode spots. In order to burn the cathode as evenly as possible the pulse length has to be long enough to allow the cathode spots generated around the trigger wire to move to the outer edge of the cathode before the pulse current is extinguished. The time required for the spots to reach the cathode edge, varies with cathode material and decreases with increasing arc current. For metal cathodes pulse lengths of approximately 0.5 ms were sufficient for peak arc currents of 3.5 kA. Because spots on graphite targets move much more slowly that spots on metal cathodes, a maximum pulse length of approximately 2 ms would be required for spots to reach the edge of a 5cm-diameter graphite cathode. Since this would require a lot of unutilised capacity in the power supply when used for metal targets, we decided instead to use smaller cathodes in the case of graphite to achieve optimum cathode consumption.

The proportion of ion current in the cathodic arc plasma plume is usually about 10% of the arc current at any time [10]. This figure does fluctuate somewhat and it has been observed [11] that the proportion of ion current for a particular arc current is highest at the start of a pulse, i.e. during the pulse rise time. This would indicate that, for a particular duty cycle, one should try to maximise the pulse frequency, i.e. sacrifice pulse length for increased frequency. Limits on pulse frequency are set primarily by the tolerance of the circuit components to heating. Our power supply is designed to have a maximum pulsing frequency of 10 Hz but it may not be possible to achieve this at the high end of the 1 to 5 kA current range.

Circuit simulations show that circuit losses will be proportional to the pulse frequency and inversely proportional to the resonant frequency. With a given L and C and hence resonant frequency, we can adjust the charge voltage to vary the peak current and the crowbar time to vary the pulse length. Our prototype dissipated only 400 W when running at maximum capacity of the wiring.

4. A Study of Cathode Spot Behaviour

In order to optimise the erosion profile of the cathode surface, we carried out a detailed study of the motion of the cathode spots in the prototype source. We used a high speed CCD camera operating with exposures down to 1 μs to photograph the light emissions from cathode spots moving on the surface of the cathode under a variety of operating conditions. The cathode materials studied included carbon, aluminium, titanium and copper.

Fig. 6 shows CCD images of arcs on carbon cathodes using the two different power supplies. Fig. 6 (a) shows an arc using the simple 12 mF capacitor bank, corresponding to the current profile in Fig. 4 (a). A large amount of arcing is observed in the central regions of the image around the trigger electrode. Conversely, the arcing in Fig. 6 (b) is not concentrated in the central regions and is more evenly spread over the cathode surface. This is due to the rising profile of the current pulse in Fig. 6 (b). As a consequence the cathode is eroded more evenly and results in better utilisation of the cathode material.

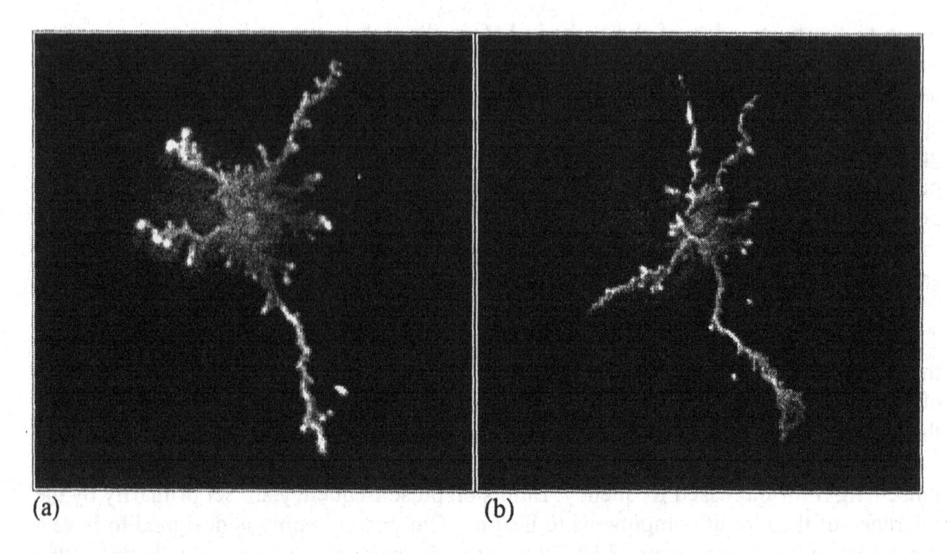

Figure 6: CCD images of carbon arcs: (a) using a 12mF capacitor bank charged to 300V; (b) using the resonant LC circuit described above, charged to 300V, crowbarred at 1ms. Exposure times are (a) 2ms and (b) 1ms. Image sizes are 5cm by 5cm.

Whilst the erosion profile is improved using the rising current pulse the erosion is complicated by the influence of the cathode spots on one another. Cathode spots repel one another due to the magnetic fields produced by the high current density transmitted through the plasma at the spot locations. When the spots are close together at the beginning of the arc pulse the force of repulsion is large. The repulsive force is also dependent on the current in the arc spot and this changes during the pulse. There is thus a complicated interplay between the current and the inter-spot distance which determines the rate at which the spots move out from the ignition electrode. Further optimisation of the current profile for efficient utilisation of the cathode material will require careful consideration and further study of these competing effects.

Two distinct types of spot behaviours were seen on all of the cathode materials examined. These spot behaviours correspond to the so-called type 1 and type 2 spots. Type 1 spots are associated with short lifetimes, low currents and high velocities. They are generally attributed to the evaporation and ionisation of contamination on the cathode surface. Type 2 spots, attributed to the evaporation and ionisation of the cathode material itself, exhibit longer lifetimes at one site, higher currents and lower velocities [12, 13]. Type 1 spots occurred on all newly inserted cathodes. After some time the spot behaviour switched sharply (within ten pulses) and permanently to type 2 behaviour. If the cathode was removed from vacuum and then reinserted, the process was repeated, i.e. type 1 spot behaviour was observed until another transition to type 2 behaviour occurred. In the deposition of multilayered films it will be important to ensure that only type 2 spots contribute to the plasma reaching the substrate.

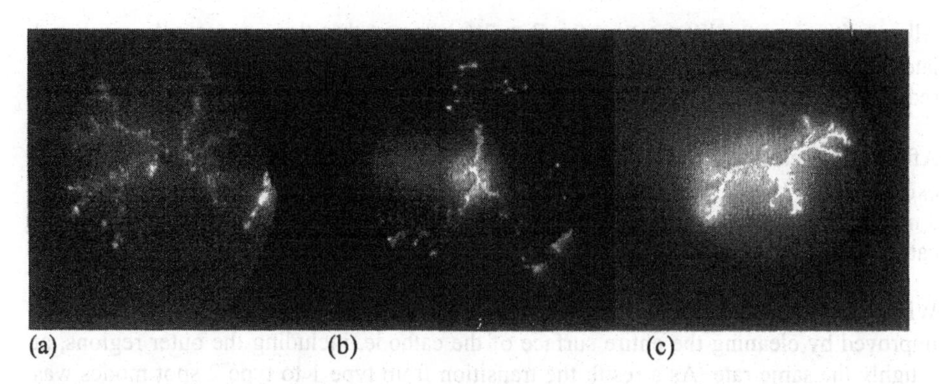

Figure 7. Comparison of different spot types on aluminium: (a) a type 1 arc (exposure time 100μs during a pulse occurring soon after the cathode was inserted into vacuum); (b) type 1 and 2 arcs occurring during a single pulse (exposure time 500 μs taken during the transition from type 1 to type 2 behaviour); and (c) a type 2 arc (exposure time 2000μs taken during a pulse occurring well after the transition). The power supply was the simple 12mF capacitor bank charged to 110V.

Fig.`7 shows examples of different types of cathode spots on an aluminium cathode. The power supply used was the simple 12mF capacitor bank referred to in the previous section, charged to 110V. The current profile was identical to that in Fig. 4 (a) except the peak current was reduced. Arc spots were initiated at the centre of the 5cm diameter circular cathode and repelled each other toward the edge of the disc. Fig. 7 (a) shows a 100μs exposure of type 1 spots burning on a newly inserted cathode. Type 1 spots exhibit comparatively high velocity and as a result the spots have reached the edge of the cathode well before the end of the 2 ms current pulse. The remaining energy of the arc pulse is concentrated into a high current discharge at the edge of the cathode associated with the formation of an anode spot. This can be observed in the photograph on the right hand side of the cathode. As a result, the plasma is contaminated due to erosion of the anode. During the type 1 mode of operation there appear to be some type 2 spots occasionally nucleated, seen as bright regions in Fig. 7 (a). The type 2 mode spots are brighter, partly because they are more slowly moving. These type 2 nuclei do not propagate, but have limited viability in view of the lower resistance path provided by type 1 operation if it is possible. Only when essentially all opportunity for type 1 operation is removed, is the type 2 mode dominant.

After approximately 50 arc pulses, depending on the cathode material and its prior exposure to contaminants, a transition occurs from type 1 and type 2 spots. The rate of this transition is dependent on the power source used. As discussed previously, the LC oscillating power supply design was used to achieve more even coverage of the cathode surface. When the simple capacitor bank power supply was used, the outer reaches of the cathode were not as effectively eroded as the inner region near to the trigger electrode. As a result, type 2 spots first occurred in the regions close to the ignition point as these areas became clean of the surface contaminants responsible for type 1 behaviour. When the moving type 2 spots reached the still contaminated areas further out they converted back to type 1 spots. An example of this is seen in Fig. 7 (b).

All conditions were the same as for Fig. 7 (a) except that the current pulse occurred later in the sequence of pulses and the exposure time was 500μs. The camera gain was reduced to avoid saturation by the high brightness type 1 component of the arc.

After a further ten or so arc pulses, the arc traces were observed to be composed exclusively of type 2 spots. Fig. 7 (c) shows a 2000 μs exposure of a type 2 arc. Having a much lower velocity, type 2 spots are still well contained within the boundary of the cathode after the entire 2 ms pulse duration.

When the optimised power supply was used, the conditioning of the cathode was improved by cleaning the entire surface of the cathode, including the outer regions, at roughly the same rate. As a result the transition from type 1 to type 2 spot modes was more abrupt, occurring within a few arc pulses. A gas analysis trace from the mass spectrometer in the vacuum chamber during the operation of the pulsed arc source (see Fig. 8) shows a sharp change in the gas species present. This corresponds exactly to the observed transition from type 1 to type 2 behaviour. From the trace it appears that hydrocarbon fragments and CO_2 are higher during the type 1 phase than during the type 2 phase, while oxygen and nitrogen are highest in the type 2 phase. This is consistent with the understanding that spots in the type 1 mode are burning a contamination layer (often of hydrocarbon) off the cathode surface, while type 2 spots are those ablating a clean cathode surface.

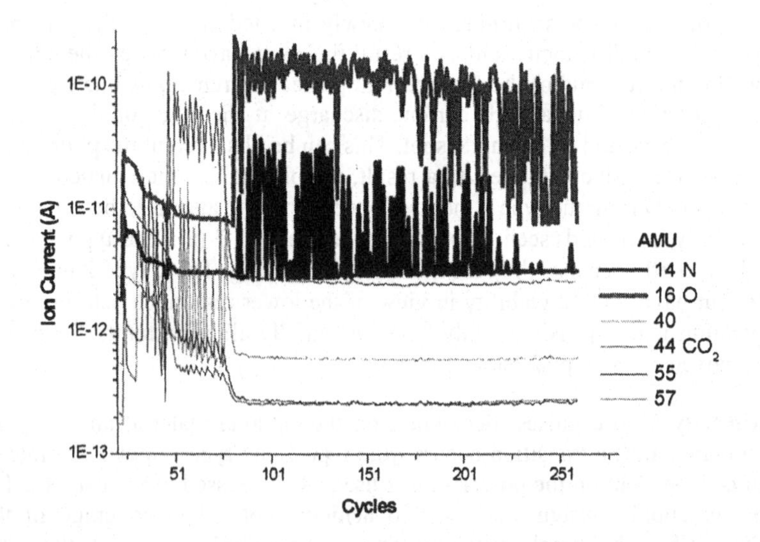

Figure 8. Mass-spectrometer trace taken during operation of a newly inserted cathode. The heavy lines are AMU 16 and 14 from top to bottom (most probably oxygen and nitrogen respectively). The light lines from top to bottom are AMU 44, 40, 55 and 57. AMU 44 could be CO_2 while the others are most likely to be hydrocarbon fragments. The pulse frequency was approximately 0.5 Hz.

For the production of well-defined multilayer films on the nanometer scale, reproducibility of the deposition rate per pulse is necessary. In our system we can achieve deposition rates of around 0.25 nm per pulse. Seimroth et al [6] have optimised their system to deposit films at up to 0.75 nm per pulse. Given that we want to fabricate nanolaminates with periodicities down to a few nanometers, this corresponds a few tens of pulses per layer. If there is a large variability in the deposition rate per pulse then the quality of the multilayer is compromised, particularly if only a few pulses are used for the deposition of each layer. For low deposition-rate sources this is less critical since the variation in the deposition rate can be averaged out over many pulses.

The influence of the spot type on the erosion rate of the cathode is therefore of great importance. As presented above, it is imperative that the cathode is always properly conditioned when a new cathode has been introduced into the system. This also brings into question the effect of the pulse frequency on the composition of the plasma produced. As pointed out by Yushkov and Anders [14], a low pulse frequency allows the formation of layers of adsorbed gases on the surface of the cathode between pulses. This gas layer subsequently affects the operation of the arc by influencing the properties of the arc spot. It also leads to incorporation of ionised gas from the adsorbed layer in the plasma.

When the multilayer consists of a ceramic portion by incorporation of a reactive gas in the deposition process the combination of background gas pressure and pulse frequency will be important. The formation time of gaseous layers on the cathode surface decreases with gas pressure, thus the recontamination rate of the cathode will increase with increasing pressure. It has been shown that the erosion rate of an aluminium cathode is strongly influenced by increases in nitrogen pressure above 10 mPa and above 1 Pa for a copper cathode [15]. This is attributed to adsorbed gases on the cathode surface influencing the nature of the arc processes. A certain minimum repetition rate will be required in order to preserve type 2 behaviour in substantial pressures of background reactive gas.

5. Conclusion

We have designed a dual source pulsed filtered cathodic arc system to deposit multilayered thin film coatings with tailored interfaces and studied the behaviour of cathode spots in the system under a variety of conditions. The pulsed mode of cathodic arc operation offers substantial advantages over the continuous cathodic arc for the deposition of multilayer films. The important characteristics and operating parameters of the pulsed arc for this application will be:

1. Controlled rates of nonreactive deposition in vacuum are possible due to the reproducible arc current pulse profile which in turn leads to a reproducible quantity of material ablated during each pulse, providing that the arc is operating on a clean cathode (i.e. in type 2 mode).

2. Controlled rates of deposition and controlled stoichiometry should also be achievable during reactive deposition processes when a background gas is used. The

rate and stoichiometry will however dependent strongly on the repetition rate and the pressure of the background gas. Ideally, the repetition rates should be chosen so that type 2 operation is maintained throughout. This will involve using a certain minimum pulsing frequency which will depend on pressure of the background gas, as well as on the type of gas and the cathode material being used.

3. The time dependence of the arc current pulse is important in giving the optimum performance. A pulse profile in which the arc current increases with time and is cut-off at a maximum value is preferred for a centre triggered cathode because it gives a sharp easily identified transition to the desired type 2 mode of operation and gives a more uniform rate of erosion of the cathode surface.

Further work will be required to find the optimum operating conditions for depositing multilayered films with any particular set of constituents. The effect of arc current and background gas pressure on the charge state distribution (CSD) of the plasma ions should also be monitored to understand how the impact energy of ions will be affected by substrate bias.

References:

1. J. Xu, M. Kamiko, Y. Zhou, R. Yamato, G. Li, M. Gu, *J. Appl. Phys.* **89**, 3674 (2001).
2. U. Helmersson, S. Todorova, S.A. Barnett, J.-E. Sundgren, L.C. Markert, J.E. Green, *J. Appl. Phys.* **62**, 481 (1987).
3. M.W. Thompson, *Phil. Mag.* **18**, 377 (1968).
4. A. Anders, *Surf. Coat. Technol.* **93**, 158 (1997).
5. M.M.M. Bilek, D.R. McKenzie, R.N. Tarrant, S.H.N. Lim and D.G. McCulloch, *Surf. Coat. Technol.* (2002) in press.
6. P.Siemroth T. Schulke T. Witke, *Surf. Coat. Technol.* **68/69**, 314 (1994).
7. B. Juttner, *J. Phys. D: Appl. Phys.* **34**, R103 (2001).
8. T. Witke, T. Schuelke, B. Schultrich, P. Siemroth, J. Vetter, *Surf. Coat. Technol.* **126**, 81 (2000).
9. E.M. Oks, A. Anders, I.G. Brown, M.R. Dickinson, R.A. MacGill, *IEEE Trans. on Plasma Sci.* **24**, 1174 (1996).
10. C.W. Kimblin, *J. Appl. Phys.* **44**, 3074 (1973).
11. H. Fuchs, K. Keutel, H. Mecke, Chr. Edelmann, *Surf. Coat. Technol.* **116-119**, 963 (1999).
12. J.M. Lafferty (1980) *Vacuum Arcs: Theory and Application,* John Wiley & Sons, New York.
13. S.Anders, A.Anders, *IEEE Trans. on Plasma Sci.* **19**, 20 (1991).
14. G.Y. Yushkov, ., A. Anders, *IEEE Trans. on Plasma Sci.* **26**, 220 (1998).
15. S. Anders, B. Juttner, *IEEE Trans. on Plasma Sci.* **19**, 705 (1991).

IMPLANTATION OF STEEL FROM MEVVA ION SOURCE WITH BRONZE CATHODE

Z. Werner[a], J. Piekoszewski[a,b], R. Grötzschel[c], W. Szymczyk[a]
[a]*Andrzej Soltan Institute for Nuclear Studies, 05-400 Swierk/Otwock, Poland*
[b]*Institute of Nuclear Chemistry and Technology, Dorodna 16, 03-145 Warszawa, Poland*
[c]*Forschungszentrum Rossendorf, Institut für Ionenstrahlphysik und Materialforschung, e.V. Postfach 510119, D-01314 Dresden, Germany*

Abstract

Bronze is known as an excellent material for bushes co-operating with steel shafts because of its low friction coefficient and low wear rate. Therefore, an improvement of the tribological properties can be expected after implanting steel with the bronze constituents (Cu and Sn). Steel samples were implanted in a MEVVA-type implanter equipped with a bronze cathode and operated at 75 kV. To determine the retained doses of the implanted elements, the samples were analyzed by RBS technique using 1.4 MeV He ions. The results show that the retained vs. implanted dose dependence saturates at implanted doses of the order of 10^{17} cm^{-2}, due to the sputtering effects. The maximum retained doses obtained for Cu and Sn amount to $5.4 x 10^{16} cm^{-2}$ and $1.2 x 10^{15} cm^{-2}$, respectively.

1. Introduction

Introduction of MEVVA type ion sources made it possible to form ion beams composed of ions of the working gas and ions of the cathode material. In this way sub-surface layers of oxides [1] or nitrides [2] of the implanted metal can be formed. However, possibilities of forming multi-element ion beam when the source cathode is made of a suitable alloy have not been explored until now. In practical applications such possibilities may be of high value, since the properties of the alloys are often much different than these of their constituents, as regards hardness, friction coefficient, resistance to oxidation etc. Thus one can expect that enhancing the content of an alloy components in the surface layer of a solid may be more beneficial than using these components separately.

Considering application of an alloy cathode in the MEVVA source, one has to take into account the question of how the cathode composition is transferred to the implanted material, and which mechanisms affect this process. The aim of the present study was to examine this problem on the example of bronze electrode. Because of its low friction coefficient and low wear rate bronze is known as an excellent material for bushes co-

E. Oks and I. Brown (eds.),
Emerging Applications of Vacuum-Arc-Produced Plasma, Ion and Electron Beams, 187–190.

operating with steel shafts. Therefore, an improvement of the tribological properties can be expected after implanting steel with the bronze constituents (Cu and Sn).

2.Experimental

20 mm dia. disks of NC-6 carbon tool steel (C 1.3-1.45%, Si 0.15-0.40%, Mn 0.4-0.6%, Cr 1.3-1.65%, V 0.1-0.25%, Fe-balance) quenched and tempered to 64 HRC and polished to a mirror finish were implanted with a direct beam of MEVVA-type source [3] equipped with the B-101 bronze cathode (Cu 89%wt, Sn 9%wt) and operated at 75 kV. The implantation equipment was manufactured by HCEI, Tomsk. According to the reference data [4] the beam is composed of singly, doubly, and triply ionized Cu ions, in the ratio 28:53:18, respectively, singly or doubly ionized Sn ions in the ratio 47:53, respectively, and several percent admixture of the working gas (nitrogen). The implanted doses of ions were $1x10^{17}$, $2x10^{17}$, and $4x10^{17}cm^{-2}$ under an assumption that the average charge state in the beam corresponds to that of the major beam constituent (1.9 for Cu).

To determine the retained doses of the implanted elements, the samples were next analyzed by RBS technique using 1.4 MeV He ions.

3. Results and discussion

The interesting fragment of the RBS spectrum of a sample implanted with the highest dose (among those used in this work) is shown in Fig.1 together with spectrum of a non-implanted sample. Only weak trace of Sn can be seen; the Cu peak overlaps with the iron edge. To elucidate the Cu content, spectrum of the non-implanted sample was subtracted from that of the implanted one.

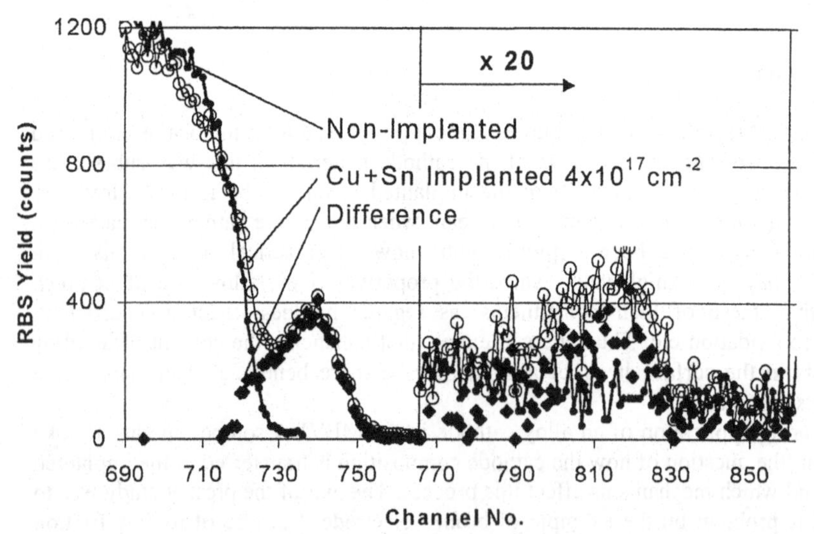

Fig.1 Fragment of RBS spectrum of a steel sample implanted with $4x10^{17}$ cm^{-2} of Cu+Sn ions from a MEVVA-type implanter equipped with the B-101 bronze electrode.

The retained doses estimated from the collected RBS spectra are shown in Fig.2 as solid circles for Cu (left scale) and solid squares for Sn (right scale) together with theoretically calculated fits (solid lines).

The fitting procedure was as follows. Firstly, a low-dose Cu profile (produced in the absence of sputtering) was computed using the SRIM computer code. The NO SPUTTERING profile shown in Fig. 3 results from superposition of profiles calculated for 75 keV, 150 keV, and 225 keV Cu ion energy, taken with the weights given by the proportion of 1+, 2+ and 3+ states in the beam, as given above. Since sputtering effects must be taken into consideration at the doses used in our experiments, the profile was further processed with another program calculating these effects. The value of sputtering coefficient Y for Cu was taken as 6, i.e. the mid-value between the values for Ar^+ and Kr^+ for 80 keV, calculated from the sputtering theory [5]. The Cu profile obtained this way is also shown in Fig.3 together with the SRIM profile.

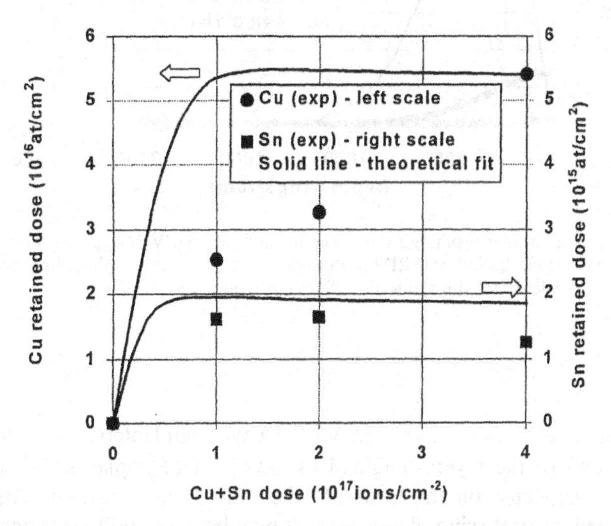

Fig. 2. Retained doses estimated from the collected RBS spectra. Solid circles: Cu data (left scale); solid squares: Sn data (right scale); solid lines: calculated fits

The calculated retained Cu dose shown in Fig.2 agrees with the experimental data only for the highest dose. An agreement at the lower-dose data would require unrealistically high value of Y coefficient. The experimental retained dose steadily increases with the Cu fluence, in contrast with the theoretical predictions. Thus we must conclude that some other processes, apart from sputtering, are involved in Cu retention in iron.

As regards Sn we assume the following model. Atomic composition of the B101 bronze is Sn 6%, Cu 94%. If ion beam is formed in the same proportions, there are 16 Cu ions per one Sn ion in the beam striking the target. Thus, the sputtering process is dominated by Cu ions, even if Sn ions have a higher sputtering coefficient. Therefore, when calculating the Sn profile and the Sn retained dose with sputtering accounted for, we assume that the Sn ions have sputtering coefficient equal to 96 (16 Cu ions times Y_{Cu}). Dependence of the measured retained Sn dose on the implanted Cu+Sn dose is shown in Fig. 2. The Sn profile calculated at the above assumptions is shown in Fig. 3. A reasonable agreement with the experimental data, much better than that for Cu, can be seen in Fig. 2.

It is also worthwhile to note that at the lowest dose the proportions between the retained Cu and Sn doses correspond to the proportions in the source cathode. For higher doses, the Sn concentration saturates at low value and the Cu concentration steadily increases.

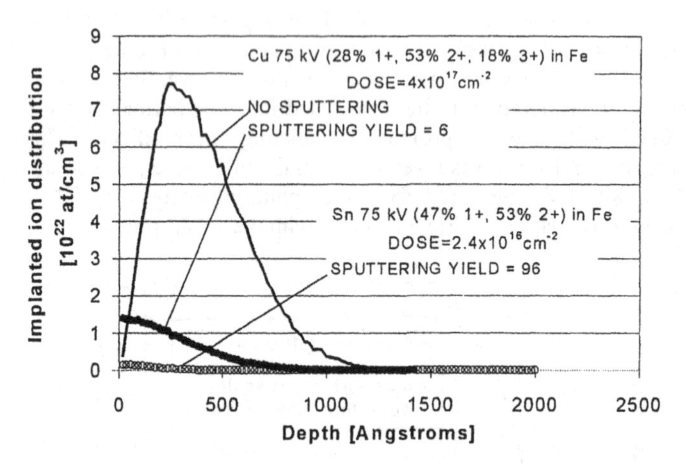

Fig. 3 Calculated Cu and Sn profiles resulting from implantation from a MEVVA-type implanter equipped with the B-101 bronze electrode. Solid line: SRIM profile calculated without sputtering. Lines with circles: the Cu and the Sn profiles calculated with sputtering effects taken into account.

4. Conclusions

In case multi-element cathodes are used in MEEVA type implanters, and high doses are applied, composition of the layers obtained as the result of implantation significantly deviates from that expected on the basis of the cathode composition. This effect is evidently dominated by sputtering phenomena. It may be expected that concentration of lighter elements will be enhanced, whereas concentration of the heavier ones will be diminished in comparison to the cathode composition.

References

1. Brown I.G., Liu F., Monteiro O.R., Yu K.M., Evans P.J., Dytlewski N., Oztarhan A., Corcoran S.G., Crowson D. (1998) *Surf.&Coat. Techn.* 103-104, 293-298
2. Yu L.D., Vilaithong T., Yotsombat B., Thongtem S., Han J.G., Lee J.S. (1998) *Surf.&Coat. Techn.* 103-104, 328-333
3. Bugaev S.P, Nikolaev A.G., Oks E.M., Schanin P.M., Yushkov G.Yu., *Rev Sci Instr* (1992) 63, 2422
4. TITAN Ion Source Manual, HCEI Tomsk, Russia
5. P. Sigmund in *Topics in Applied Physics, Vol 47 Sputtering by Particle Bombardment I, R. Behrish (ed) Springer Verlag, Berlin, Heidelberg, New York, 1981*

RESISTANCE TO HIGH TEMPERATURE OXIDATION IN Si-IMPLANTED TiN COATINGS ON STEEL

Z. Werner[a], J. Piekoszewski[a,b], W. Szymczyk[a]
[a]*Andrzej Soltan Institute for Nuclear Studies, 05-400 Swierk/Otwock, Poland*
[b]*Institute of Nuclear Chemistry and Technology, Dorodna 16, 03-145 Warszawa, Poland*

Abstract

Moulds for light-alloy die casting are usually made of TiN-coated AISI H13 tool steel. During operation the moulds undergo wear caused (among others) by high-temperature corrosion (oxidation). Recently it has been shown that Ti-Si-B-N coatings prepared by reactive sputtering from multi-element Ti-Si-B targets exhibit excellent resistance to high-temperature corrosion, and that such resistance can also be obtained by ion implantation of Si and B into TiN coating on H13 steel. However, the B-ion beam cannot be produced in MEVVA-type ion sources. Therefore an attempt was made to improve the high-temperature oxidation resistance of the TiN coating by implanting Si ions alone. Thermogravimetry measurements show that $1x10^{17}$ cm^{-2} Si ions implanted from modified MEVVA-type ion source operated at 75 kV results in more than twofold reduction of the oxidation rate at 630°C.

1. Introduction

In the field of improving wear and corrosion resistance of tools or machine parts, ion implantation has usually been considered as an alternative to hard coating deposition in those cases, in which deposition is (for some reasons) undesired, or leads to inferior results as compared to those obtained by ion implantation. However, ion implantation has also been examined as a method for improving the surface properties of objects with ceramic protective coatings [1-8]. Most of the referenced investigations is focused on improvement of wear characteristics of the TiN, TiC, and TiAlN coatings by N and C implantations. Other studied alternatives include implantation of Ni and Ti to improve the lifetime of Al_2O_3-coated WC-Co cutting tools [5], and implantation of boron to improve the tribologic properties of TiN and TiAlN coatings [7,8] – the latter with an excellent result.

On the other hand, ion implantation was frequently used as a tool to study the effect of alloy additives on the corrosion properties, in particular on high-temperature corrosion of various steels and alloys [9-11]. Range of implanted species was much wider in these studies: it included such exotic elements like yttrium and cerium.

E. Oks and I. Brown (eds.),
Emerging Applications of Vacuum-Arc-Produced Plasma, Ion and Electron Beams, 191–195.

The problem of influence of ion implantation on high-temperature corrosion resistance of ceramic protective coatings has been examined only very recently [12]. This point may be of significant practical importance, since in many cases ceramic protective coating are exposed to high temperatures (like in light-alloy moulds or as cutting edge of the cutting tool operated in no-lubricant conditions). Moulds for light-alloy die casting are usually made of TiN-coated AISI H13 tool steel. During operation such moulds undergo wear caused by mechanical action of light alloy injected under a high pressure, corrosive attack of the alloy, thermal cycling, and high-temperature corrosion (oxidation). Therefore, the coating may deteriorate as a result of oxidation rather than direct wear, loosing that way its hardness and mechanical strength.

At this point it is worthwhile to recall the results obtained by Shtansky et al. on high-temperature resistance of reactively magnetron-sputtered protective coatings of Ti-Si-B-N composition obtained using TiB_2+(20wt%)Si targets and Ar+N_2 gas mixtures [13]. The coating composition, as deduced from AES profiles was: 35 at% of Ti, 35 at% of B, 20 at% of N, 10at% of Si. Exposed to air at 550°C for 2h such coatings exhibited reduction of the oxidation rate by a factor of at least 10 with respect to the TiN coating. The measured microhardness value H_v was 3800 as compared to a value of 1600 obtained for TiN and the wear rate was reduced by a factor of about 1.3 with respect to TiN.

The results obtained by Werner et al.[12] are also encouraging. By using RBS analysis of the composition of oxidized TiN layers they were able to show that the oxidation rate at 630°C was reduced by a factor of at least 6 as a result of Si+B implantation to a dose of $2x10^{17}cm^{-2}$.

To implement such result in industrial practice, the cost of the implantation process must be low, and comparable with the cost of other surface treatments. This can be achieved only if a high-intensity ion sources are used, such as MEVVA type ones. Since B ions cannot be produced in such sources, it seems worthwhile to investigate effects of implanting Si ions alone on the high-temperature corrosion properties of TiN coatings. Another motivation for such study is that Si implantation is known to produce substantial increase in high-temperature corrosion resistance in Ti-based alloys [11].

2.Experimental

Samples were cut from H13 steel as rectangular plates of dimensions 25x8x1 mm, quenched and tempered to about 52 HRC. The latter treatment is standard in preparing parts of light-alloy moulds. Subsequently the samples were coated with about 2μm thick TiN layer by commercial Arc PVD coating process and implanted on both flat sides with a direct beam of MEVVA-type source [14] with a silicon cathode, operated at 75 kV. The implantation equipment was manufactured be HCEI, Tomsk. The beam is composed of 75keV (63%), 150keV (35%), and 225keV(2%) Si ions and several percent admixture of the working gas (nitrogen). The implanted doses were $2x10^{16}$, $5x10^{16}$, and $1x10^{17}cm^{-2}$.

Non-implanted and implanted samples were next subjected to thermogravimetric (TG) analysis, in which the increments of the sample mass were recorded during 1h isothermal annealing at 630°C in a stream of standard technical air. The temperature of annealing was selected on the basis of temperature scan in the 500-1200°C range, performed to determine the temperature, at which a 1 h annealing leads to oxidation of about 1 μm thick layer in untreated TiN.

3. Results and discussion

Results of thermogravimetric tests are presented in Fig.1 as mass increments vs. time of the isothermal annealing. The data indicate that the oxidation rate is markedly reduced as a result of ion implantation, and that the effect increases with the ion dose. The reduction factor attains a value of 2.36 for the highest applied dose. Keeping in mind that the TiN coating reduces the oxidation rate with respect to the non-coated steel by a factor of 2.5 [12], the reduction in the oxidation rate resulting from TiN coating and Si implantation jointly amounts to about 6.

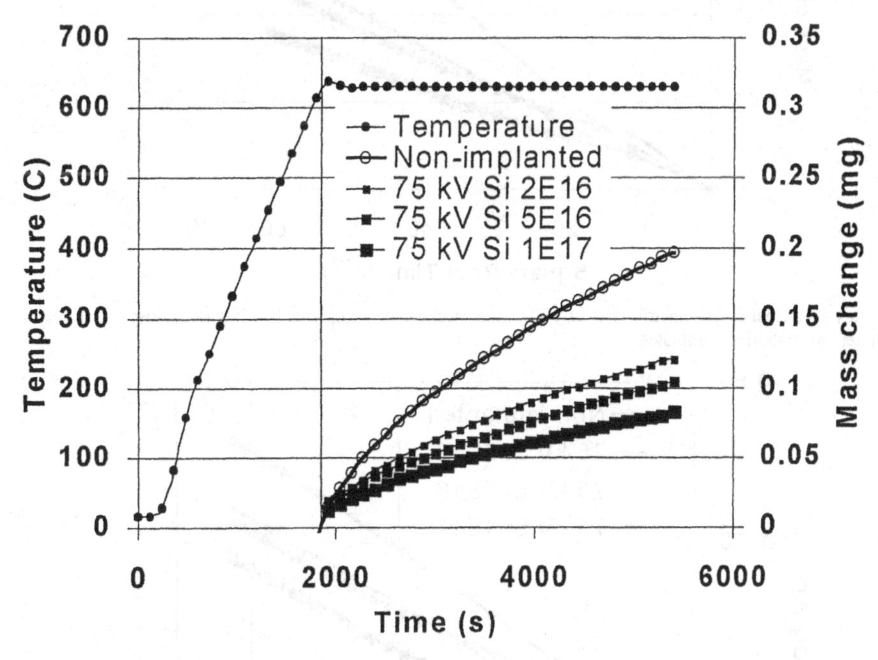

Fig.1 Results of thermogravimetric tests of TiN-coated steel samples implanted with various doses of Si ions

To study the kinetics of the oxidation process, the data of Fig.1 have been re-plotted in Fig.2 in parabolic coordinates. TG curves for implanted samples follow quite well straight lines, thus indicating that we deal with a typical oxidation process. The rate of this process is limited by diffusion of oxygen through the growing oxide layer. It apparently slows down as the number of silicon atoms acting as oxygen traps increases. However, the TG curve for non-implanted sample does not follow this parabolic behaviour. When the data are again re-plotted in logarithmic coordinates (Fig.3), it can be noticed that the non-implanted sample follows very precisely the $t^{2/3}$ dependence. The theory of oxidation process [15] predicts that the oxidation process should start with a linear section (~t), corresponding to the reaction rate limited process, when the oxide thickness is small. This section is then followed by a parabolic section ($\sim t^{1/2}$)

corresponding to the diffusion-limited process. It seems that oxidation observed in non-implanted sample corresponds to an intermediate region, when both, reaction rate and diffusion, affect the course of the process.

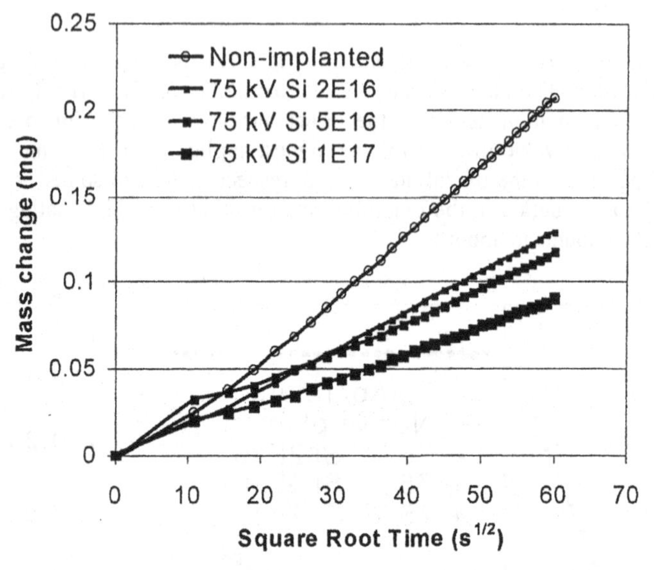

Fig.2 Results of thermogravimetric tests of TiN-coated steel samples implanted with various doses of Si ions re-plotted in parabolic coordinates.

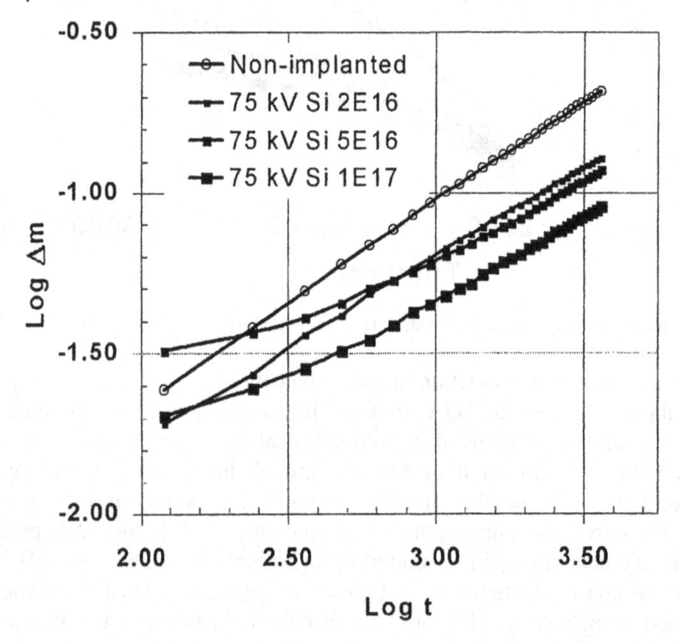

Fig.3 Results of thermogravimetric tests of TiN-coated steel samples implanted with various doses of Si ions re-plotted in logarithmic coordinates.

4. Conclusions

Si implantation leads to substantial improvement of high-temperature oxidation resistance of TiN protective coatings and can be applied instead of a more complicated and expensive B+Si implantation. It seems that the presence of implanted Si atoms slows down diffusion of oxygen through the growing oxide layer, making the process of oxidation the diffusion-limited one. Since high-intensity Si ion beams can be easily obtained in the existing MEVVA type ion sources, the obtained results offer a perspective of practical implementation.

References

1. Andreyev M., Anishchik V.(2001) Vacuum 63:541-544
2. Roos J.R., Celis J.P., Franck M., Pattyn H. (1991) Surf & Coat Techn 45:89-98
3. Sansom D., Viviente J.L., Alonso F., Ugarte J.J., Onate J.I..(1996) Surf & Coat Techn 84:519-523
4. Liu L.J., Sood D.K., Manory R.R., Zhou W. (1995) Surf & Coat Techn 71:159-166
5. Treglio J.R., Tian A., Perry A.J. (1993) Surf & Coat Techn 62:438-442
6. Kulkarni A.V., Mate N., Kanetkar S.M., Ogale S.B., Wagh B.G. (1992) Surf & Coat Techn 54/55:508-515
7. Dmitrova V.I. (1998) Vacuum 49:199-204
8. Zhu Y-C., Fujita K., Iwamoto N., Nagasaka N., Kataoka T. Proc SMMIB 2001 Conference (to be published)
9. Bennet M.J., Tuson A.T. (1989) Mat Sci & Eng A116:79-87
10. Stroosnijder M.F. (1997) Surf Eng 13:323-329
11. Haanappel V.A.C., Stroosnijder M.F. (1999) Surf Eng 15:119-125
12. Werner Z., Piekoszewski J., Grötzschel R., Richter E., Szymczyk W. (2002) Proc. Symposium, ION'2002, Kazimierz/n. Wisłą, Poland, 10-13.06.2002, to be published in Vacuum
13. Shtansky D.V., Levashov E.A., Sheveiko A.N., Moore J.J. (1998) J Mat Synth & Proc 6:61
14. Bugaev S.P, Nikolaev A.G., Oks E.M., Schanin P.M., Yushkov G.Yu. (1992) Rev Sci Instr 63:2422
15. See e.g. Deal B.E., Grove A.S. (1965) J.Appl. Phys.36:3770

VACUUM ARC DEPOSITED DLC BASED COATINGS

O.R. MONTEIRO*, M.P. DELPLANCKE-OGLETREE**
*Lawrence Berkeley National Laboratory University of California
Berkeley, California 94720, USA
**Industrial Chemistry, Université Libre de Bruxelles,
Brussels 1050, Belgium

Abstract

The great interest in the use of diamond-like carbon (DLC) films as a coating material is justified by the superior wear resistance and hardness, chemical inertness, and very low friction coefficients of these coatings. Vacuum arc deposition is well suited to prepare superhard films with high sp^3/sp^2 ratios. However, the high level of internal stresses originating during growth prevents the deposition of thick films, and their hardness makes it difficult for DLC layers to comply with substrate deformations. In order to overcome these limitations, different approaches are possible. Multilayer structures are one means to maintain the surface mechanical properties of the DLC while relieving the internal stresses. Another possibility is to dope the DLC films in order to reduce the internal stress and to stabilize the desirable sp^3 bonds to higher temperatures. At higher doses of dopants, the formation of nanocrystals is possible and the properties of the coatings change drastically. All these approaches were investigated on films prepared by cathodic arc and a synthesis of the results is presented here.

1. Introduction

In response to the challenges of implementing diamond-like carbon (DLC) into several applications, different strategies have been developed in order to enhance their performances. Addition of alloying elements such as refractory metals [1], silicon [2-4], nitrogen and fluorine was extensively investigated. Incorporation of such elements typically leads to improvements in some properties at the expense of others.
The formation and properties of nanocomposite, functionally graded layers and multilayers involving DLC has also been the subject of extensive work [5-6]. The results are very promising [7].

Stresses can reach values higher than 10 GPa in pure non-hydrogenated DLC [8], and about 3 GPa in hydrogenated DLC [2]. If there is a general acceptance in the literature on the effect of alloying elements on the reduction of internal stress, the results

E. Oks and I. Brown (eds.),
Emerging Applications of Vacuum-Arc-Produced Plasma, Ion and Electron Beams, 197–203.

concerning other properties are not as generally accepted. For example, reports on the influence of the silicon concentration on the hardness of the DLC doped films are contradictory [2, 9, 10]. Most of the work on silicon additions to DLC has involved hydrogenated films. The effect of tungsten and titanium additions on the mechanical is highly dependent on the concentration of the alloying element that induces changes in the structure of the films.

Many parameters such as proportion of the different layers, the structure of the inferfaces control the properties of multilayers. A comparison of the properties of coatings prepared by different techniques is thus rather difficult.

The work presented here focuses on non-hydrogenated DLC based coatings prepared by filtered cathodic arc and will briefly summarized results on mechanical properties and structure of silicon and tungsten doped layers, TiC-DLC nanocomposites and TiC-DLC multilayers.

2. Experimental procedure

2.1. DEPOSITION

The various coatings were prepared using two filtered cathodic plasma sources and a pulsed bias voltage applied to the substrate. Depending on the desired structure of the coatings (doped DLC, multilayers, nanocomposite), the two sources were triggered simultaneously or successively. In all cases, one source was fitted with a graphite cathode and the other with the addition element cathode: Silicon, titanium or tungsten. For the doped and nanocomposite layers, the composition was controlled by varying the duration of the pulses in each source. During each plasma pulse a fraction of a monolayer was deposited. No layering was observed in these conditions. For the multilayers, the carbon source was working by itself for the deposition of the pure DLC components. The titanium source was added for the synthesis of the TiC_x component of the multilayers.

A pulsed bias voltage was applied to the substrate during each arc pulse. The pulse duration of the sources varied between 0.5 and 10 ms. The arc frequency was 1 Hz and pulsed bias voltage of –100 V was applied to the substrate with alternating on-periods of 2 μs and off-periods of 6 μs. To obtain a diffuse interface between the substrate and the coatings or between the different layers composing the coatings, a higher bias voltage of –2000 V was imposed for a short period before switching to –100 V.

2.2. CHEMICAL AND STRUCTURAL CHARACTERIZATION

Quantitative chemical composition of the deposited coatings was obtained by Rutherford Backscattering Spectroscopy using a 1.8 keV He^+ beam. X-Ray photoelectron spectroscopy (XPS) and Auger Electron Spectroscopy (AES) were used to characterized the bonding in the films and the depth distribution. Transmission electron microscopy was used to characterize the morphology and structure of the films, TEM was carried out in a Phillips CM200 with a PEELS system and a Topcon 002B with point resolution of 0.19 nm at 200 kV.

2. 3. MECHANICAL AND TRIBOLOGICAL CHARACTERIZATION

Hardness and elastic modulus of the films were determined by nanoindentation using a Hysitron Picoindenter. The hardness was evaluated from the residual impression of the indenter whereas the modulus was evaluated from the slope of the load-unload curve at the beginning of the unloading process [11]. A Hysitron Triboscope was used to determined the friction coefficient. It was determined using a single-pass scratch test. Ball on disk wear tests were performed to evaluate the macroscopic wear resistance of the layers using a CSEM apparatus. The test were performed in 50% relative humidity, with a 2N load, a 6mm in diameter alumina ball at 0.2 m/s speed.

3. Results and discussions

3..1. SILICON DOPED AND TUNGSTEN DOPED LAYERS

The Si-content of the DLC:Si films used in this work are 3 at%, 5 at% and 6 at% as determined by RBS. The W-content of the DLC-W films varied between 1 and 10 at%. The film composition and structure result from a balance between the indicidant ion currents from the two plasma sources and the sputtering rates of the individual elements due to the incident energetic Si^{n+} or W^{n+} and C^{+} ions.The mean charge state of the Si^{n+} ions generated by the cathodic arc source is 1.4 while the mean chage state of the W^{n+} ions is higher than 3. This is extremely important for the mean energy transferred to the growing film. It is dependent on the proportion of the elements incorporated in the films all other conditions remaining constant. It is known that the properties of undoped DLC depends strongly on the energetics of the deposition. However, for low ratio of addition element in the mixed plasma stream the final effect of the difference in energy transfer should be relatively small especially in the case of silicon.

For both system, the films are totally amorphous in the range of concentration considered here. The sp^3 content increases with the incorporation of silicon in the films [10], the highest ratio being obtained for 3 at%. The hardness values of DLC/Si films obtained at several loads are shown in figure 1. The values given in this figure are not corrected for the deformation of the diamond indenter. The "real" hardness is probably slightly lower than the values that are quoted. The hardness of undoped DLC films prepared in similar conditions is of the order of 72 GPa. The addition of silicon induces a hardness decrease but it is accompanied by a large decrease of internal stress.
A similar behavior is observed for W:DLC films as shown in figure 2 a and b. The addition of 1at% of tungsten reduces the stress by a factor of three but decreases the hardness only by 15%. In both systems, the best compromise between reduced stresses and hardness is obtained at low concentration of the addition element. The weaker Si-C

and W-C bonds could be responsible of the reduction of the stress and of the hardness. These bonds also stabilize the sp^3 configuration of the carbon atoms. The very strong effect of tungsten could also be related at high concentration to the modification of the sp^3/sp^2 ratio resulting from the large energy input per W atom.

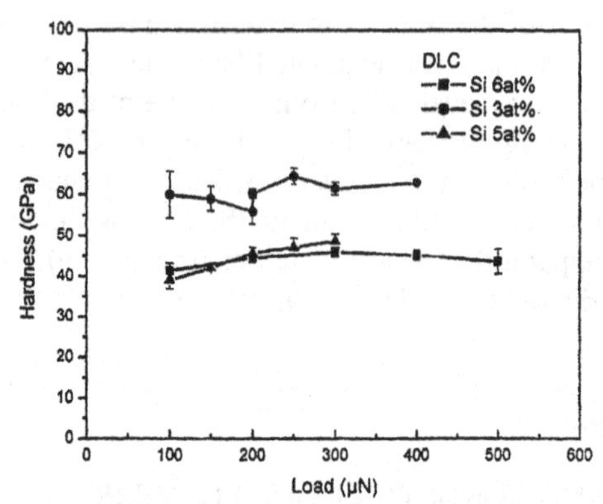

Figure 1: Hardness of the DLC:Si films obtained at several loads

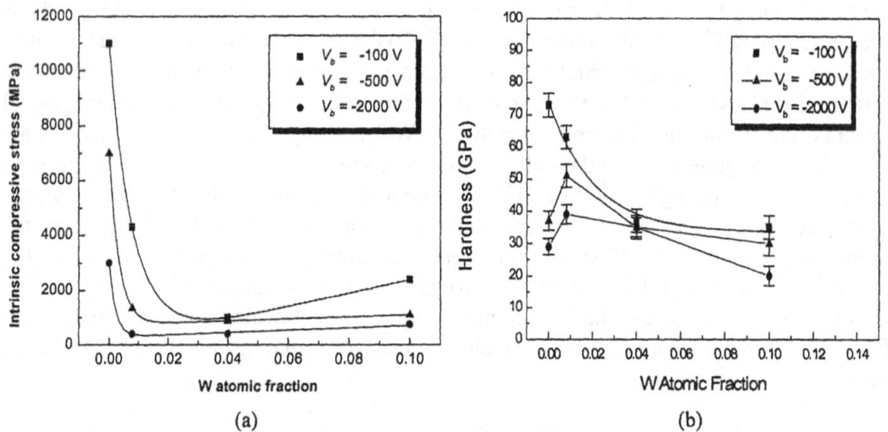

(a) (b)

Figure 2: variation of (a) internal stress and (b) hardness with the W-content in DLC:Z films

3.2. TiC_x-DLC MULTILAYERS

Figure 3 presents the structure of three multilayers based on TiC_x and pure DlC components. Each interface was deposited at a bias voltage of –2000V while the bulk is deposited at –100 V for ensuring a good adhesion between the different components of the film. Figure 4 gives the hardness and elastic modulus of the same layers at different loads. None of these values was corrected for the deformation of the indenter tip. This deformation is responsible of the abnormally high value of the registered hardness The decrease of hardness as a function of load reflects the hardness variation with indentation depth. The contribution of TiC_x layers, having a lower hardness than pure DLC, increases with the indentation depth and thus reduces the total hardness of the system.

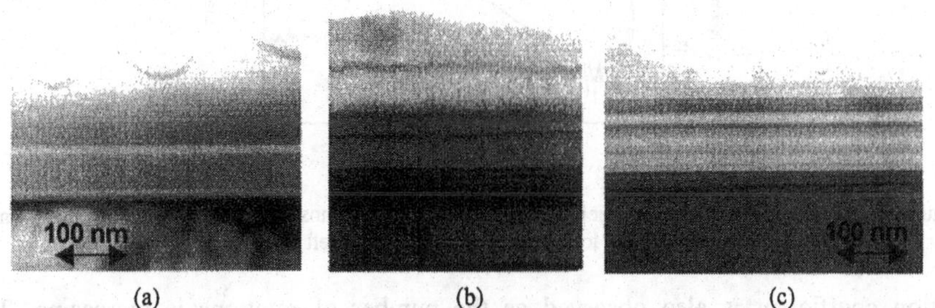

(a) (b) (c)

Figure 3: TEM of multilayers (a) DLC-TiC_x-DLC-Si, (b) DLC-TiC_x-DLC-TiC_x-DLC-Si and (c) DLC-TiC_x-DLC-TiC_x-DLC-TiC_x-DLC-Si

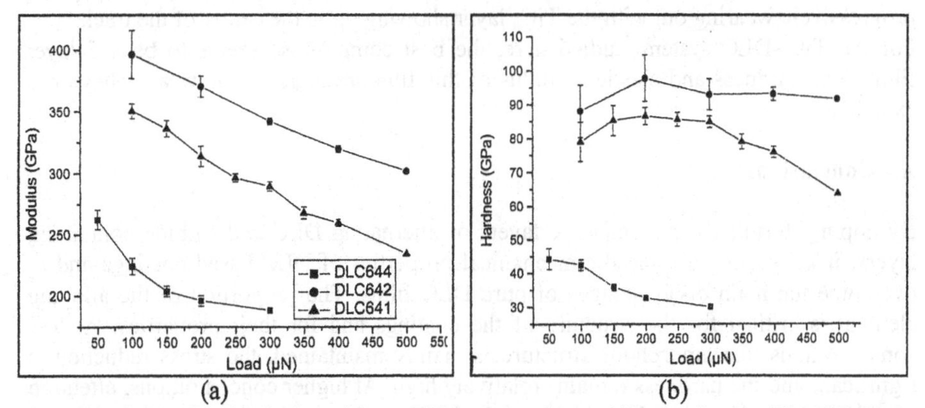

(a) (b)

Figure 4: variation of the (a) elastic modulus and (b) hardness of the multilayers: triangle = DLC-TiC_x-DLC-Si, dots = DLC-TiC_x-DLC-TiC_x-DLC-Si and squares = DLC-TiC_x- DLC-TiC_x-DLC-TiC_x-DLC-Si

The elastic modulus follows a similar trend. The strong reduction of the hardness and modulus for the 7 layers film is due to the high proportion of interfaces in this film. The interfaces are soft because the sp^2 content is high for films deposited with a –2000V bias. The friction coefficient measured during a single-pass nanoscratch reflects the penetration of the indenter in the different layers when the load is increased as shown on figure 5. The step noticed when the normal load reach a value of 100 µN is probably due to the penetration of the indenter in the second layer. This layer is softer and has another composition. The friction coefficient of TiC is higher than for DLC. The step increase in partially due to the change of contact nature but also to an increase of the plowing contribution to the friction coefficient.

The value of the friction coefficient obtained at the nanometric scale is similar to those measured during macroscopic wear tests. In those tests, a step-like variation of the

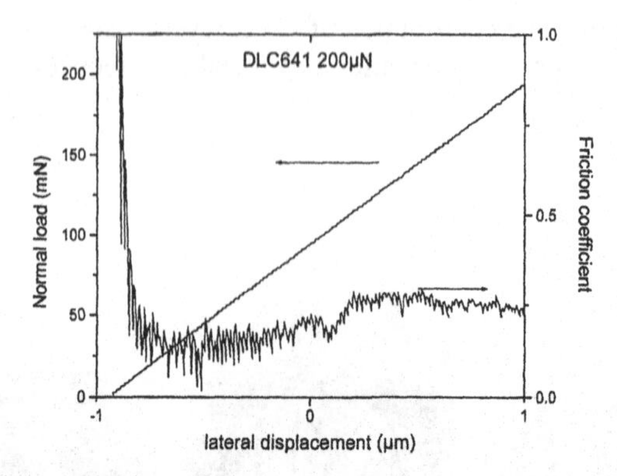

Figure 5: variation of the friction coefficient during a single-pass nanoscratch in the DLC-TiC_x-DlC film . The normal load increases linearly from left to right

friction coefficient is also observed as the number of rotations is increasing. The contact of the alumina ball with the TiC_x layer is probably also responsible of this sudden increase. In this conditions, the normal load is constant but the top DLC layer is progressively wearing out with the TiC_x layer showing up in the center of the track.
For the TiC_x-DLC system studied here, the best compromise seems to be a 5 layer films. The hardness and elastic modulus of this film are high and its wear behavior is excellent.

4. Conclusions

By doping, forming nanocomposite layers or alternating DLC and carbide containing layers, it is possible to tune the mechanical properties of DLC based coatings and to overcome the main disadvantages of pure DLC films. The proportion of the alloying element is critical for the structure of the coatings and for their properties. At low concentrations, the amorphous structure is mainly maintained, the stress reduction is significant and the hardness remains relatively high. At higher concentrations, attention must be paid to the mean charge state of the alloying element because it can modify the energy transferred to the growing film and thus indirectly change the sp^3 to sp^2 ratio. The alloying element can form a separate phase giving rise to a nanocomposite structure. The parameters controlling the mechanical properties in those films differ from those observed in the doped layers.
The proportion of "soft" interface and the relative thickness of the DLC and TiC_x layers are the key factors for the mechanical properties of the multilayers.

Acknowledgements

The authors would like to thank the Computer Mechanics Laboratory at the University of California, Berkeley and the F.N.R.S for continuous support

References

1. Delplancke-Ogletree, M.P. and Monteiro, O.R. (1998), *Surface and Coatings Technology* **108-109**, 484-488
2. Baia-Neto, A.L., Santos, R.A., Freire Jr, F.L., Camargo Jr, S.S., Carius, R., Finger, F., and Beyer, Z. (1997), *Thin Solid Films* **293**, 206-211
3. Shi, J.R., Shi, X., Sun, Z., Liu, E., Yang, H.S., Cheah, L.K., and Jin, X.Z. (1999), *J. Phys.:Condens. Mater.* **11**, 198-203
4. Lee, C.S., Lee, K.R., Eun, K.Y., Yoon, K.H., and Han, J.H. (2002), *Diamond and Related Materials* **11**, 198-203
5. Voevodin, A.A., and Zabinski, J.S. (1998), *Diamond and Related Materials* **7**, 463-467
6. Miyoshi, K., Pohlchuck, B., Street, K.W., Zabinski, J.S., Sanders, J.H., Voevodin, A.A., and Wu, R.L.C. (1999), *Wear* **115-119**, 65-73
7. Delplancke-Ogletree; M.P., and Monteiro (1997), *J. Vac. Sci. Technol. A* **15(4)**, 1943-1950
8. Monteiro, O.R., Ager III, J.Z., Lee, D.H., Yu Lo, R., Walter, K.C., and Natasi, M. (2000), *J. Appl. Phys.* **88(5)**, 2395-2399
9. Zhao, J.F., Lemoine, P., Liu, Z.H., Quinn, J.P., Maguire, P., and McLaughlin, J.A. (2001), *Diamond and Related Materials* **10**, 1070-1075
10. Monteiro, O.R., and Delplancke-Ogletree, M.P. (2002), *Surface and Coatings Technology*, accepted
11. Oliver, W.C., and Pharr, G.M. (1992), *J. Materials Res.* **7(6)**, 1564-1583

APPLICATIONS OF VACUUM ARC PLASMAS TO NEUROSCIENCE

I.G. Brown, O.R. Monteiro, E.A. Blakely, K.A. Bjornstad,
J.E. Galvin and S. Sangyuenyongpipat
Lawrence Berkeley National Laboratory
Berkeley, California 94720 USA

Abstract. To understand how large systems of neurons communicate, we need to develop, among other things, methods for growing patterned networks of large numbers of neurons. Success with this challenge will be important to our understanding of how the brain works, as well as to the development of novel kinds of computer architecture that may parallel the organization of the brain. Large *in vitro* networks could show, for example, the emergence of stable patterns of activity and could lead to an understanding of how groups of neurons learn after repeated stimulation. We have investigated the use of metal ion implantation using a vacuum arc ion source, and plasma deposition with a filtered vacuum arc system, as a means of forming regions of selective neuronal attachment on surfaces. Lithographic masks created by treating surfaces with ion species that enhance or inhibit neuronal cell attachment allow subsequent proliferation and/or differentiation of the neurons to form desired patterns. Plasma deposition of optically transparent, electrically conducting, ultra-thin metal films can also be used to form electrodes for extra-cellular electrical stimulation of neurons. Substrates tested in our work were primarily glass microscope slides; some of the experiments made use of simple masks to form patterns of ion beam or plasma deposition treated regions. PC-12 rat neurons were then cultured on the treated substrates coated with Type I Collagen, and the growth and differentiation was monitored. Particularly good results were obtained, for example, for the case of plasma deposition of carbon to form a diamond-like carbon film of thickness about one hundred Angstroms. Neuron proliferation and the elaboration of dendrites and axons after the addition of nerve growth factor both showed excellent contrast, with prolific growth and differentiation on the treated surfaces and very low growth on the untreated surfaces. Here we describe our preliminary investigations, and summarize the results to date.

1. Introduction

The study of the functional unit of the nervous system, the neuron, has been an active field of investigation for many years, both at the single-cell level, *in vivo* and *in vitro,* and at the level of large numbers of interconnected neurons, for example within the human brain [1]. The behavior of individual neurons has been studied using microelectrodes to monitor the electrical signals ("action potentials") generated within the neuron and along its dendrites (the branch-like arms that carry signals toward the neuron cell body where they are processed) and axons (the long "tail" that carries the neuron output signal to other cells). One can think of the single-cell electrical behavior

E. Oks and I. Brown (eds.),
Emerging Applications of Vacuum-Arc-Produced Plasma, Ion and Electron Beams, 205–211.

as the performance at the "device level" [2], and at this level much is known. At the "system level", however, much less is known – we know very little about how large numbers of neurons communicate among themselves. There has been good progress made in the growth of neuron cultures *in vitro*. The neurons grow, extend dendrites and axons, form synapses, and create neural networks. In order to systematically explore the electrical characteristics of large numbers of associating neurons, however, we need first to develop techniques for forming 2-dimensional patterned arrays of large numbers of neurons. All of the parameters of the patterning should be under the control of and determined by the experimenter, including the geometry of the pattern, the line width, and the pattern size (number and density of neurons). The subsequent step is a daunting step indeed – we need then to discover and develop methods for monitoring the electrical activity throughout the array. Methods for monitoring the activity of small numbers of neurons have been developed, mostly making use of extracellular recording of the action potentials with extracellular microcircuit electrode arrays. Note that the use of chronic intercellular (impaled) microelectrodes (micron-size electrodes penetrating the cell wall) is not feasible because of cell death (the cell dies within a few hours of electrode insertion), and because the method cannot be extended to large arrays. It has not been possible up to now to adequately detect the spatial geography and temporal history of the action potentials in large neuronal arrays. Several approaches to patterning have been explored [3], including mechanical fabrication of troughs and ridges [4], laser micromachining [5], surface photochemical methods [6], photoresist methods, among others. These methods work and have been used to grow neural arrays. Here we describe some exploratory work that we have carried out investigating the suitability of vacuum-arc-plasma based methods of surface modification as a tool for forming large patterned neuronal arrays.

2. Experimental

Neuron cultures can be grown on glass or tissue culture plastic substrates coated with Type I Collagen. In the work described here we used ordinary glass microscope slides of dimension 1" x 3". In the first part of the work, our goal was to explore the effects on neuron growth of either (i) metal ion implantation into the glass, or (ii) plasma deposition onto the glass surface. Implantation was done using the vacuum arc ion source ("Mevva V")-based ion implantation system that has been fully described elsewhere [7-9]. We note that implantation into insulating target materials, such as glass, is readily accomplished in this kind of system without any charge build-up problems; we have shown in prior work [10] that the Mevva-produced high current metal ion beam provides self-neutralization of insulating targets via the background cold electron sea formed by the ion beam itself, alleviating the need for any additional electron source for target charge neutralization. The implants were done at a relatively low energy of about 10 to 30 keV, to doses from 5×10^{14} to 1×10^{16} cm^{-2}, and with several different ion species including C, Mg, Al, Ti and Ta. Plasma deposition was done using the filtered vacuum arc system that has been previously described [9,11,12]. Briefly, metal plasma is created by a repetitively-pulsed vacuum-arc plasma gun, with 5 msec pulses at a rate of 1 pps. The plasma was filtered with a 90° magnetic duct, and the glass substrate was mounted on a grounded holder (no pulse-biasing in this work) positioned about 10 cm from the duct exit. Films were in this way formed on the glass microscope slides, of thickness in the approximate range 30 – 300 Å. Ion species used, and thus the kinds of film materials formed, included C, Mg, Ti, Pd, Ta, Ir, Pt and Au. By doing the depositions at a somewhat elevated background pressure it is straightforward to form metal oxides, and thus we also made films of aluminum oxide, titanium oxide and tantalum oxide. A characteristic feature of vacuum-arc-produced plasmas is the relatively high directed energy with which the ions are formed, in the approximate range 20 to 150 eV depending on the ion species [13]. The film deposition is thus an energetic deposition, and for the case of carbon this results in the film material formed being a high quality, hydrogen-free, diamond-like carbon (DLC) [14,15], not amorphous carbon or graphite. As described below, we found that the carbon films were particularly advantageous for enhanced neuron growth.

We obtained PC-12 neurons from the American Type Culture Collection (Manassas, VA). The PC-12 cell-line was derived from a transplantable rat pheochromocytoma from the adrenal gland. The cells are grown in RPMI 1640 media with 2 gm/L glucose (Invitrogen), 10% heat-inactivated horse serum (Invitrogen), 5% fetal bovine serum (HyClone), 2 mM L-glutamine, 1.5 g/L sodium bicarbonate, pen strep at 37^0C, 7.5% CO_2 on Type I Collagen coated Biocoat^{TM} (Becton Dickinson) plastic 100 mm petri plates. Stock cultures were fed every three days with 2/3rds fresh media, and subcultured every 9 days with a 1:4 cell split ratio. Nerve Growth Factor (NGF) 2.5S (Invitrogen) was added to cell densities at concentrations of 50 ng/ml. On a collagen-coated substrate, neurite elongation proceeds at an average rate of ~50 μm/day for at least 10 days. After 2 weeks of NGF exposure, the cultures generate a dense mat of neuritic processes. Generally, at least 90-95% of the cells in the cultures produce neurites.

PC-12 cells were inoculated onto pre-cleaned, plasma-deposited DLC-coated, Type I Collagen-coated sterile glass slides at 1 x 10^5 cells/ml. Cells were allowed to adhere to the slide in a 7.5% CO_2 incubator at 37^0 C, for 3 hours, and then gently flooded with growth media. Cell growth was monitored by phase light microscopy. Cells were photographed with a digital Spot Camera on a Nikon TMS scope using the Spot Advanced software, and printed using Adobe Photo Shop. After 3-6 days of cell growth, NGF was added to the media at 50 ng/ml. After the addition of NGF, cell division stops and differentiation begins. PC-12 cells double every 96 hours. Cultures were visually monitored daily and images captured every other day, up to 1.5 months after initiation of the cultures.

3. Results

Neurons grew on all the processed substrates, but there was a wide variation observed in the total number of attached cells and their morphology. Under identical neuron growth conditions for each substrate surface tested, the neuronal cell density attained in the cultures was found to vary over many orders of magnitude for the various processing methods investigated. That is, we can state unequivocally that processing of the glass substrate in the ways described above does indeed provide a means of controlling the neuron growth. Ion implantation, for the entire range of parameters explored, was found to be universally negative in its effect – the growth rate and the culture cell density at all times during the growth were both very poor compared to the case of plasma deposition. For example, glass slides that had been ion implanted with carbon under a range of conditions, including at particularly low energy (10 keV) so as to form a carbon profile close to the glass surface, and at relatively high dose (1 x 10^{16} cm^{-2}) so as to increase the surface carbon concentration, also yielded negative results. We quickly abandoned further exploration of ion implantation as a tool for enhanced neuron growth.

Plasma deposition, on the other hand, was seen to provide significantly enhanced neuron growth for some kinds of film materials (plasma deposition species). We found that the metals provided a generally positive growth enhancement and that all of the metal oxides were generally negative in their effect. The single film material that stood out as providing vastly enhanced growth was carbon, which as descxribed above is deposited in the form of hydrogen-free diamond-like carbon, or DLC. We therefore chose to investigate neuron growth on carbon surfaces in more detail. Variation of DLC film thickness indicated that a film thickness of about 100 – 150 Å was near optimum. For thinner films, the neuron "contrast ratio" – ratio of neuron growth density on the DLC-coated region to density on the non-DLC-coated region – was less, and thicker films tended to delaminate from the substrate.

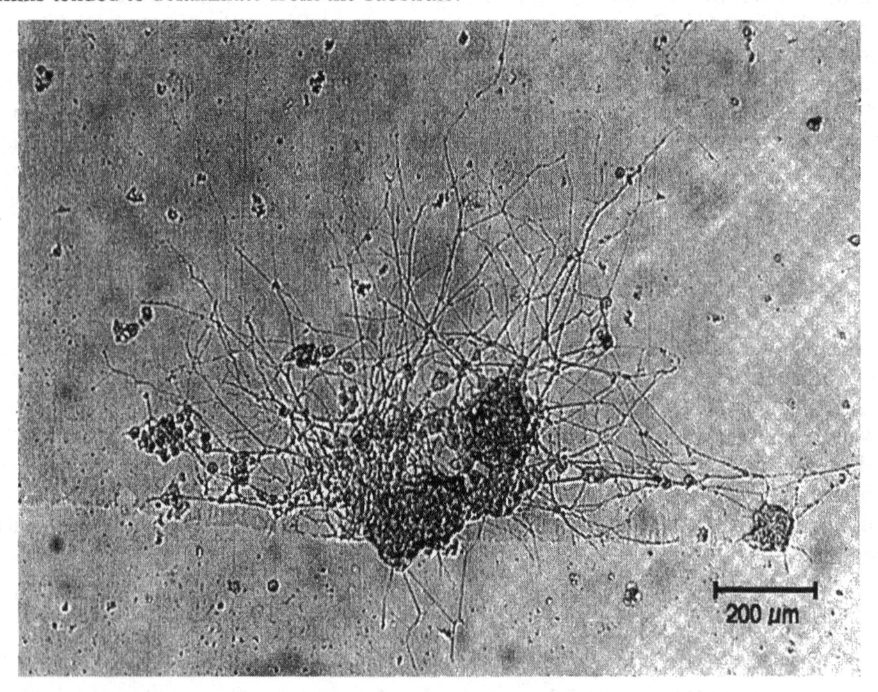

Figure 1. Selective neuron growth on DLC-coated substrate.

Neuron growth after 15 days on a glass slide onto which a 100 Å thick film of DLC was deposited. The DLC region can be seen as a slightly darker region occupying the upper 75% of the whole region viewed; there was no DLC coating on the lower part of the image. The whole slide was coated with Type I Collagen.

The photograph in Figure 1 shows clearly how neurons grew preferentially on a DLC coated substrate. One can see that (i) neuron growth is healthy on the upper DLC-coated region, with virtually no growth on the lower uncoated region, (ii) in the region of good growth, the DLC region, neurons grow extended processes (axons and neurons), (iii) the neuron extensions show a pronounced tendency to confine their growth to the DLC region.

The results of another growth experiment are shown in Figure 2. Here the neuron density is prolific, much greater than would be chosen for a controlled experiment. But the point is made beautifully clear that the growth is limited to only the DLC-coated region. Neuron growth is on a glass substrate processed by plasma deposition of ~100 Å coating of diamond-like carbon (DLC) film. The plasma deposition was such that the lower part of each photo is the DLC-treated region, and the upper part is not DLC-treated. The entire substrate was collagen coated, and the neurons were seeded over the entire surface. The left-hand photo shows the delicate neurite growth that develops on the DLC-treated region; the right-hand photo shows that the neuron growth in the DLC-treated region continues to a dense and prolific neuron density. These results indicate that neurons grow selectively on the lower DLC-treated regions and not on the upper untreated regions. The contrast (ratio of neuron density in the treated region to neuron density in the untreated region) is very high, and neuron growth in the treated region is healthy.

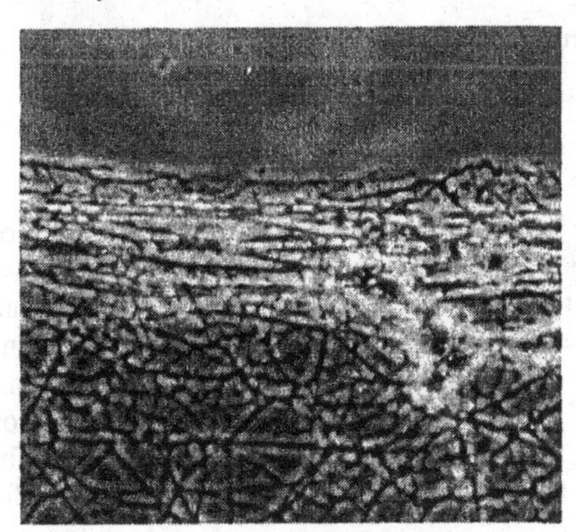

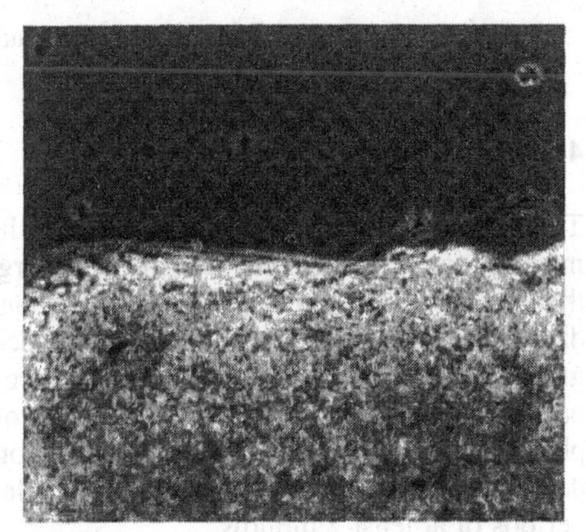

Figure 2. Selective PC-12 rat neuron growth on collagen-coated, DLC-plasma-processed surface.

The lower party of each photograph shown was DLC coated, with the upper part not coated; the substrate was collagen coated, and neurons were then seeded over the entire surface. A delicate neurite growth develops on the DLC-treated region (left-hand photo), which develops into a dense and prolific neuron field (right-hand photo). (Scale: the width of each photograph is several hundred microns).

The results of our first attempt at neuron patterning are shown in Figure 3. Neuron growth is on a glass substrate processed by plasma deposition of ~150Å diamond-like carbon (DLC) film. Prior to deposition, "LBNL" was written on the glass slide using a fine marker pen, and then the DLC deposition was carried out. After DLC deposition, the ink was removed with alcohol, thus leaving "LBNL" patterned in negative in the DLC film. The slide was then coated with Type I Collagen and seeded with PC-12 rat neurons. The neurons were allowed to grow for 3 days, at which point NGF (Nerve Growth Factor) was added. The micrographs shown in Fig. 3 were taken after a growth period of 6 days after initiation of the cultures.

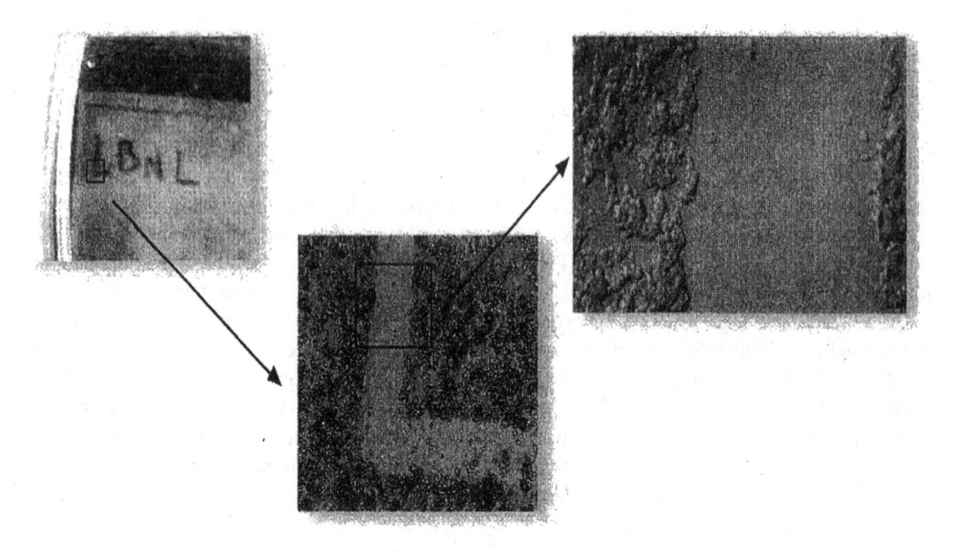

Figure 3. Patterned Growth of Neurons

4. Conclusion

The work described here has demonstrated the suitability of filtered-vacuum-arc deposition methods for forming patterned arrays of large numbers of live neurons. We have shown that energetic plasma deposition of carbon to form an ultra-thin layer of amorphous diamond on the substrate surface provides a means for selective neuron attachment, growth, and differentiation on that surface. The optimal DLC thickness for neuronal patterning is ~100 Å to 150 Å. The neuron growth contrast (ratio of neuron density on plasma-treated regions to neuron density on untreated regions) can be very high. The deposited carbon films are strong and stable, and remain intact on the glass substrate for periods of at least 2 months.

References

1. D. Vaudry, P.J.S. Stork, P. Lazarovici, L.E. Eiden, Signaling pathways for PC12 cell differentiation: Making the right connections, *Science* **296**, 1648-1649 (2002).
2. C. Wilkinson and A. Curtis, Networks of living cells, *Physics World*, **12**(9), 45-48 (1999).
3. D.A. Stenger and T.M. McKenna, *Enabling Technologies for Cultured Neural Networks*, Academic Press, San Diego (1994).
4. C. Miller, H. Shanks, A. Witt, G. Rutkowski and S. Mallapragada, Oriented Schwann cell growth on micropatterned biodegradable polymer substrates, *Biomaterials* **22**, 1263-1259 (2001).
5. J.M. Corey, B.C. Wheeler and G.J. Brewer, Compliance of hippocampal neurons to patterned substrate networks, *J. Neurosci. Res.* **30**, 300-307 (1991).
6. J.J. Hickman, S.K. Bhatia, J.N. Quong, P. Shoen, D.A. Stenger, C.J. Pike and C.W. Cotman, Rational pattern design for in vitro cellular networks using surface photochemistry, *J. Vac. Sci. Tech.* **A12**(3), 607-616 (1994).
7. I.G. Brown, Vacuum arc ion sources, *Rev. Sci. Instrum.* **65**(10), 3061-3081 (1994).
8. I.G. Brown, M.R. Dickinson, J.E. Galvin, X. Godechot and R.A. MacGill, Versatile high current metal ion implantation facility, *Surf. Coat. Technol.* **51**, 529-533 (1992).
9. I.G. Brown, A. Anders, M.R. Dickinson, R.A. MacGill and O.R. Monteiro, recent advances in surface processing with metal plasma and ion beams, *Surf. Coat. Technol.* **112**, 271-277 (1999).

10. F. Liu, O.R. Monteiro, K.M. Yu and I.G. Brown, Self-neutralized ion implantation into insulators, *Nucl. Instrum. Meth.* **B132**, 188-192 (1997).
11. S. Anders, A. Anders and I. Brown, Macroparticle-free thin films produced by an efficient vacuum arc deposition technique, *J. Appl. Phys.* **74**(6), 4239-4241 (1993).
12. See also R.L. Boxman, P.J. Martin and D.M. Sanders, editors, *Vacuum Arc Science and Technology*, Noyes, New York (1995).
13. A. Anders and G.Yu Yushkov, Ion flux from vacuum arc cathode spots in the absence and presence of a magnetic field, *J. Appl. Phys.* **91**, 4824-4832 (2002).
14. G.M. Pharr, D.L. Callahan, S.D. McAdams, T.Y. Tsui, S. Anders, A. Anders, J.W. Ager, I.G. Brown, C.S. Bhatia, S.R.P. Silva and J. Robertson, Hardness, elastic modulus, and structure of very hard carbon films produced by cathodic-arc deposition, *Appl. Phys. Lett.* **68**, 779-781 (1996).
15. O.R. Monteiro, Plasma synthesis of hard materials with energetic ions, *Nucl. Instrum. Meth. Phys. Res.* **B148**, 12-16 (1999).

CONCERNING REGULARITIES OF PARTICLE'S DRIVING IN POTENTIAL FIELDS (ON EXAMPLE OF ELECTRON'S MOVEMENT IN ELECTRICAL FIELD WITH DISTRIBUTED POTENTIAL).

V.I. Fedulov
Institute of Power Engineering
of Automation of Academy
of Sciences of Uzbekistan,
Academgorodok, Tashkent,
Uzbekistan

Abstract. The aim of this work is the research of the trajectory of movement and of the characteristics of radiation of an electrons' flow in the flat structures with braking electrical fields, created by an accelerating electrode and an electrode with distributed potential. The analytical expressions for definition of the electron's movement trajectory in the suggesting structure, the conditions of the absolute electron's energy transfer, the conditions of the cyclic closed trajectory's appearance and of the linear cyclic vibrations of the electron in the researching electrical field were found. The conditions, where the electron entering the structure could stay in it or could fly from it were defined, i.e. the conditions of completed or partial absorption of its initial kinetic energy. Under research of movement's trajectories it was discovered, that the appearance of sections with subzero resistance («the tunnel effect») is possible in structures with distributed potential, and, probably, it is possible only in structures with distributed potential. Is shown, that in potential fields with a distributed potential does not exist uniform and rectilinear movements. The analytical expressions for the definition of a difference of velocities of electron's movement in a structure, taking off from one point, but directed under an angle 90^0 rather each other are obtained. The symmetrical relative to X and Y axles electrical field with distributed potential was offered and the electron's movement in such field was investigated. The conditions, when the electron accomplished circular cyclical and rectilinear cyclical vibrations in such field were defined. The generation's process of electromagnetic radiation was investigated in structures and the analytical expressions determining the dependence of generated radiation's frequency on structure's p parameter were received depending on value of initial velocity and value of its coordinate of entry's in structure. The opportunity of optical (Cherenkov's effect) radiation's appearance in ionised solid mediums is shown in the work. Is shown, that considering an anisotropy electrical field with a distributed potential and gravitational field, in which A. A. Michelson has carried out experiences, as potential, it is easy to see, analogy in expressions for a relative modification of velocities of electrons and speeds of light in longitudinal and transversal directions.

E. Oks and I. Brown (eds.),
Emerging Applications of Vacuum-Arc-Produced Plasma, Ion and Electron Beams, 213–225.

1. Introduction.

The physics of transformation process of kinetic energy of electron's flow in electrodynamics structures represents a large scientific and practical interest. The widely known researches and the theoretical analysis of dynamics of movement and process of energy transformation of an electron flow are based on notions about interaction of an electron's flow with homogeneous electrical fields [1]. The work is devoted to research of physical laws of the movement's trajectory and the interaction of an electron's flow in the electrical field with distributed potential. The electrical field with distributed potential is formed by plane-parallel structure consisting of two electrodes, on one of which the constant (accelerating) potential is fed, and on another–the potential, which is distributed under the certain law (let us assume–under the linear law).

2. Concerning the trajectory of electron's movement in electrical field with distributed potential in plane-parallel structure.

The analysis of movement's trajectory of the electron flow as an independent task underlies in creation of physical devices and installations such as generators of electromagnetic vibrations, injectors, accelerators and etc. Moreover, this analysis plays the essential role at the analysis of physical processes in physical installations. The task consists in the description of the movement's trajectory of electrons between plane-parallel electrodes, on one of which the distributed potential feeds, let us suppose, under the linear law, and on another – the constant potential (fig. 1). In the beginning the task is reduced to the definition of an electrical field in a structure with parameters, and then to the definition of the equations of driving in it.

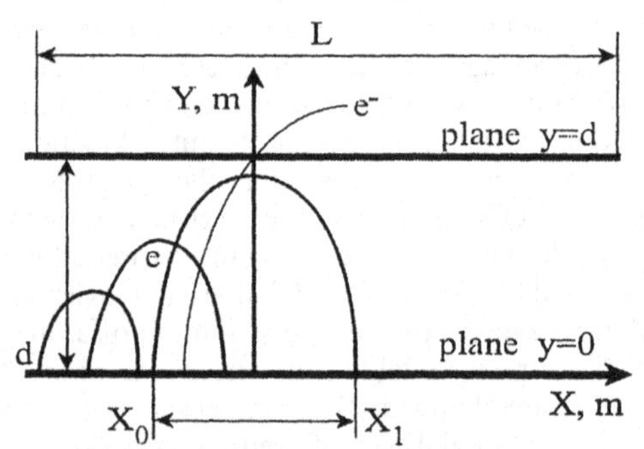

Figure 1. The configuration of the task and the orientation of the co-ordinates.

The parameters of a structure have values, which are given as $U_1(x,0)=0$ (V), $U_2(x, d)= kx$ (V), where: $U_1(x,0)$– the value of potential on planes y=0 (V), and $U_2(x, d)$– the value of potential on planes y = d (V), L- the length of a structure (m), $k=(U_{max}-U_{min})/L$– the coefficient of potential's distribution (V/m), and d – the width of structure (m). where: U_{max}, U_{min} – the value of maximal potential and minimal potential, which are applied to the ends of the electrode in the plane y = d (the electrode with distributed potential), the electrode in the plane y = 0 - the electrode with constant potential (accelerating,)= 0 (V). It is shown in [2,3] that the electrical field in the plane-parallel structure is described by the equation:

$$U=\frac{k}{d}xy \tag{1}$$

The field between the planes with the given potentials will be electrostatic and plane-parallel and satisfies to the two-dimensional Laplace's equation.

$$\frac{\partial^2 U}{\partial x^2}+\frac{\partial^2 U}{\partial y^2}=0 \tag{2}$$

The intensity of electrical field will determine from the known expression:

$$E=-\text{grad}U=-\frac{k}{d}\left(y\bar{x}_0+x\bar{y}_0\right) \tag{3}$$

Where: x_0 and y_0 – basis vectors of co-ordinates. The equations of electron's movement in the finding field will be the system:

$$\begin{cases} \ddot{x}=-\dfrac{e}{m}E_x=\alpha y \\ \ddot{y}=-\dfrac{e}{m}E_y=\alpha x \end{cases} \tag{4}$$

where: m, e, – the mass and the charge of the electron, and $\alpha=\frac{ek}{md}$ – the parameter of the system. The determination in the modulating kind has a type:

$$2x(t)=(x_0+y_0)\text{Coshpt}+\frac{V_{0x}+V_{0y}}{p}\text{Sinhpt}+(x_0-y_0)\text{Cospt}+\frac{V_{0x}-V_{0y}}{p}\text{Sinpt} \tag{5}$$

$$2y(t)=(x_0+y_0)\text{Coshpt}+\frac{V_{0x}+V_{0y}}{p}\text{Sinhpt}-(x_0-y_0)\text{Cospt}-\frac{V_{0x}-V_{0y}}{p}\text{Sinpt} \tag{6}$$

$$2Vx(t)=p(x_0+y_0)\text{Sinhpt}+(V_{0x}+V_{0y})\text{Coshpt}-p(x_0-y_0)\text{Sinpt}+(V_{0x}-V_{0y})\text{Cospt} \tag{7}$$

$$2Vy(t)=p(x_0+y_0)\text{Sinhpt}+(V_{0x}+V_{0y})\text{Coshpt}+p(x_0-y_0)\text{Sinpt}-(V_{0x}-V_{0y})\text{Cospt} \tag{8}$$

where: $p=\sqrt{\alpha}$ – the value of characteristic equations' root.

The equations (5) and (6) describe the trajectory of electron's movement in modulating kind, and (7) and (8) – its velocity. The typical trajectories of electron's movement are represented at the fig. 2,3 and 4. In all represented figures the Y axle passes through a point, where the potential on the electrode with distributed potential has equal value with the potential of accelerating electrode. Let us decide on analysis of submitted figures of the trajectory of electron's movement in the electrical field with distributed potential. At Fig. 2 the trajectories of electron's movement in structure depending on the value of coordinate of entry's point under the constant value of entry's velocity are represented.

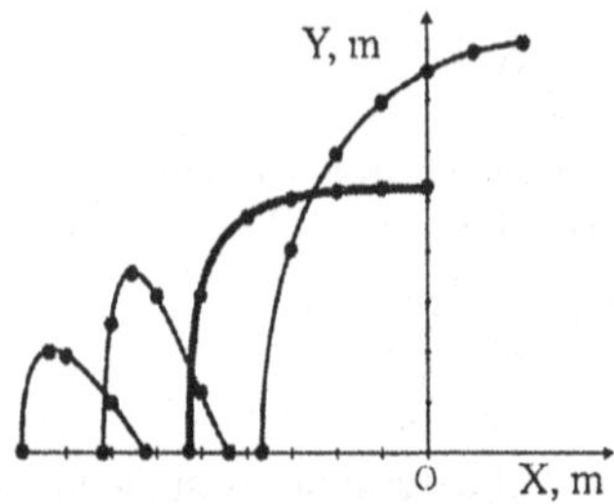

Figure 2. The trajectories of electron's movement in structure depending on the value of coordinate of entry's point under the constant value of entry's velocity

There is a border in the electrical field with distributed potential, where the electron under the same value of initial velocity depending on values of its initial velocity's constituents and also on the coordinate of its entry's point can stay in structure or can leave it. It means, that the electron loses the value of its initial velocity either completely, or partially. In other words, in such structure the transformation of electron's kinetic energy to the electromagnetic radiation can occur either completely, or partially. Let us determinate this dependence. We receive the following from the expressions (7) and (8):

$$px_0 = 2V_0 \frac{\mathrm{Sin}(pt + \mathrm{arctg}\, {}^{V_{0X}}\!/_{V_{0Y}})}{(e^{pt} + \sqrt{2}\mathrm{Sin}(pt + {}^{\pi}\!/_{4})} - (V_{0X} + V_{0Y}) \tag{9}$$

Under the t=0 the expression (5) has a kind:

$$px_0 = V_0 \mathrm{Sin}\,(\mathrm{arctg}\, {}^{V_{0X}}\!/_{V_{0Y}}) - (V_{0X} + V_{0Y}) = B \tag{10}$$

The expressions (9) and (10) functionally connect the p parameter, the initial electron's velocity V_0 and velocity's constituents V_{ox} and V_{oy} in plane-parallel structure with distributed potential with the coordinate of its entry's point x_o, on the electrode with distributed potential. Let us analyse the expression (10): Under $px_0 \leq B$ the electron will lose the initial kinetic energy completely in structure with generation of electromagnetic radiation (the kinetic energy is absorbed completely). Under these conditions the electron doesn't leave the structure, and accomplishes some movement in structure, then it comes back at electrode with accelerating potential again. Under this movement the electron brakes at the beginning, and then it increases the kinetic energy again to value, which is determinate by the potential of accelerating electrode. There is the partial selection of energy under $px_0 \rhd B$ and the electron comes out the limits of structure (the kinetic energy is absorbed partially). If the electron enters normally in a structure ($V_{OY} = V_0, V_{OY} = 0$), the initial value of a final velocity Vf, at which the electron remains in a structure, will be defined as:

$$px_0 = -V_{0Y} = V_f \tag{11}$$

For a determinacy we shall accept this initial value for a value of a final velocity Vf, at which the electron remains in a structure. Ratio of a velocity of electron's movement in a structure (from 7 and 8) to any constant of a velocity, is admissible to a value of a maximum velocity Vf, at which the electron remains in a structure will be defined by expression:

$$\frac{V(t)}{V_f} = \sqrt{\frac{V_0^2}{V_f^2} + \frac{V_{0y}p^2x_0t + 2V_{0x}V_{0y}p^2t^2}{V_f^2} + \cdots} \tag{12}$$

At equal values initial and maximum velocities and in view of a condition (11) expressions (14) are accepted by an aspect:

$$\frac{V(t)}{V_f}=\sqrt{1-\frac{2V_{0y}^2}{V_f^2}pt+\frac{2V_{0x}V_{0y}}{V_f^2}p^2t^2-} \qquad (13)$$

Expression (13), (5) and (6) show, that in electrical fields with a distributed potential does not exist uniform and rectilinear motions, and the parameter p characterises a configuration of an electrical field and accordingly magnitude of interaction of an electrical field with an electron and determines rate of acceleration or braking of an electron in it. The expression (13) determines a modification of a velocity of driving of an electron in an electrical field with a distributed potential concerning a constant it of a velocity of driving (for example Vf). It similarly adopted to [1] expression, for the definition of a modification of a velocity of driving relative of constant value of light's speed with, except that instead of a velocity With is present a final value of a velocity of an entry to a structure, at which electron does not go out it. At Fig 3. the trajectories of electron's movement in structure depending on the value of entry's velocity under the constant value of coordinate of entry's point are represented.

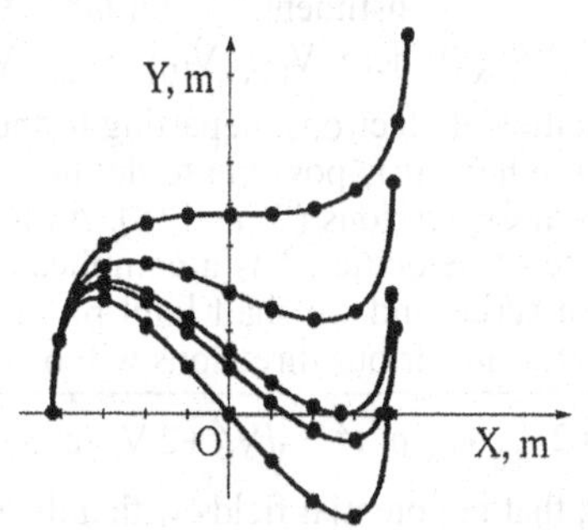

Figure 3. The trajectories of electron's movement in structure depending on the value of entry's velocity under the constant value of coordinate of entry's point

When the initial entry's velocity has diminution, on the trajectory of electron appears bend and under these conditions the electron can across the Y axle and reach the electrode with distributed potential. When we have similar conditions of electron's entry, but there are uniform electrical fields, the electron will not reach the electrode with distributed potential, and will come back to accelerating electrode. The curving of trajectory of an electron's driving determines emerging a site (zone) with negative values of conductivity in a structure. This process, probably, is similar to «tunnel» effect originating in media. And then it is possible to make the supposition, that the «tunnel» effect arises because of presence of a distributed potential and is possible (probable) only in such structures. At Fig 4 the trajectories of electron's movement in structure depending from value of entry's angle under the constant value of entry's velocity are submitted Naturally, was detected, that at driving of an electron in various direction it the velocity is various. This circumstance is peculiarity of electron's movement in such fields.

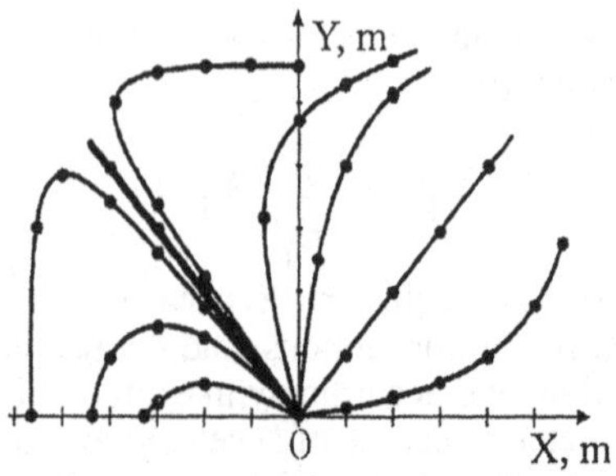

Figure 4. The trajectories of electron's movement in structure depending from value of entry's angle under the constant value of entry's velocity

Let's consider of electron's movement, taking off from one point, under an angle 90^0, rather each other and we shall define a residual between their velocities. For simplification of comprehension and without prejudice to comprehension of physics' process, let us accept that $y_0 = 0$. For the electrons departing under right angles relative to each other, for observance of velocities' equality in both directions, must carry out following proportion of values of constituents' initial velocities of electrons, namely $V_{20y}=-V_{10x}$, $V_{20y}=V_{10x}$. Where: $V_{10x}, V_{10y}, V_{20x}, \dot{V}_{20y}$ –according to values of constituents' initial velocities of electrons, departing to the different directions (under right angles relative to each other). It is possible to define a residual between velocities of electrons, proceeding from expressions (7) and (8). As the analytical expression, for the definition of a difference of velocities, has a complicated aspect, we shall take advantage of expansion it in a series and we shall limit by quadratic terms of expansion. The residual between velocities in various directions will be defined as:

$$\Delta V(t) = \sqrt{V_0^2 + 2\,V_{0y} p^2 x_0 t + 2 V_{0y} V_{0x}\, p^2 t^2} - \sqrt{V_0^2 + 2\,V_{0x} p^2 x_0 t - 2 V_{0y} V_{0x}\, p^2 t^2} \qquad (14)$$

The expressions (13) show, that in potential fields with a distributed potential of a velocity of electrons in various directions is various, and their residual can be determined with the help of expressions (13), and to their residual is equal 0 only in an initial moment. Examining analysing electrical field with distributed potential and gravitation field, where A.A. Michelson carried out his experiments, like potential fields, it is easy to see the analogy in expressions for velocities' difference of electrons and velocities' difference of light in longitudinal and transverse directions and it is becoming clearly about the difference of velocities of light in longitudinal and transverse directions in A.A. Michelson's experiments, which were carried out in gravitation and, probably, heterogeneous field. It is possible to make a conclusion that in potential fields with the distributed potential does not exist uniform and rectilinear movement, and it is possible to explain effect of a difference of velocities in various directions by anisotropy properties of such fields. The given supposition requires additional researches. Submitted at Fig. 2,3,4 trajectories of electrons flow's movement in structure, represent typical process of electron's «reflection» and «refraction» of trajectory of electrons flow's movement in analysing structure. Probably, suggesting approach can be used for analysis of interaction (for example, for analysis of interaction of processes of «reflection» and «refraction») particles with potential fields.

3. Concerning the cyclic trajectories of electron's movement in electrical field with distributed potential in plane-parallel structure.

The analytic description of process of electron's movement in structure is of scientific and practical interest, when the electron makes cyclic vibrations under the condition of completely absorption of initial kinetic energy in structure. Let us rewrite the equations (5) and (6) taking into account the conditions (11), then we shall have following:

$$2x(t)=(x_0+y_0)\mathrm{Chpt}+(x_0+y_0)\mathrm{Shpt}+(x_0-y_0)\mathrm{Cospt}+(y_o-x_0)\mathrm{Sinpt} \quad (15)$$

$$2y(t)=(x_0+y_0)\mathrm{Chpt}+(x_0+y_0)\mathrm{Shpt}-(x_0-y_0)\mathrm{Cospt}-(y_o-x_0)\mathrm{Sinpt} \quad (16)$$

Adding and subtracting (15) and (16), we shall receive

$$x(t)+y(t)=(x_0+y_0)e^{pt} \quad (17)$$

$$x(t)-y(t)=-\sqrt{2}(x_0-y_0)\mathrm{Sin}(pt-\pi/4) \quad (18)$$

Multiplying the equations (17) and (18) we shall receive the equation for description of electron's movement in the citing at the Fig. 1 structure:

$$x^2(t)-y^2(t)=\sqrt{2}\,(x_0^2-y_0^2)\,e^{pt}\,\mathrm{Sinpt}(pt-\pi/4) \quad (19)$$

The expression (19) is the analytic description of electron's movement in structure with distributed potential under the condition of completely absorption of its initial kinetic energy.

Adding and subtracting (17) and (18), we shall receive

$$2\,x(t)=(x_0+y_0)e^{pt}+\sqrt{2}(x_0-y_0)\mathrm{Sin}(pt-\pi/4) \quad (20)$$

$$2y(t)=(x_0+y_0)e^{pt}-\sqrt{2}(x_0-y_0)\mathrm{Sin}(pt-\pi/4) \quad (21)$$

Raising to the second power the expressions (20) and (21) and adding them we shall receive:

$$2[x^2(t)+y^2(t)]=(x_0+y_0)^2e^{2pt}+2(x_0-y_0)^2\mathrm{Sin}^2(pt-\pi/4) \quad (22)$$

or

$$2r^2(t)=r_0^2[e^{2pt}+2\,\mathrm{Sin}^2(pt-\pi/4)]-2x_0y_0[e^{2pt}-2\mathrm{Sin}^2(pt-\pi/4)] \quad (23)$$

Accepting value of ordinate at the point of entry $y_0=0$ as initial, the expression (23) will have a kind:

$$2r^2(t)=r_0^2[e^{2pt}+2\mathrm{Sin}^2(pt-\pi/4)] \quad (24)$$

The expression (24) is the analytic description of electron's movement in structure with distributed potential under the condition of completely absorption of its initial kinetic energy and this expression represents the circumference with changeable, depending on time, value of radius. Under t=0 we shall receive:

$$r(t=0)=r_o \quad (25)$$

If the value of r_0 coincides with the value of x_0 and the conditions (11) carry out, the electron will circumscribe closed curve. It is, probably, to say that in order to the electron has made closed trajectory, we should create symmetrical relative to X and Y axles electrical field with distributed potential. We can create such field with the help of parallel-situated electrodes with distributed negative potentials relative to the centred, and with the help of electrode with the positive potential, which is situated between them (Fig. 5). The potentials of centred and of the accelerating electrode have equal values. The picture of electrical field and closed trajectory of electron are submitted at fig. 5.

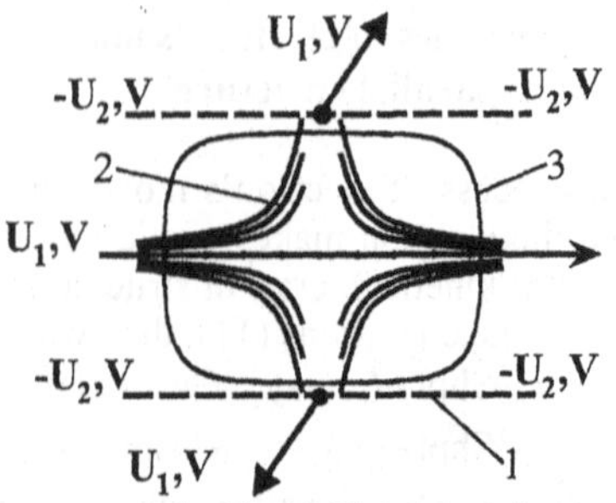

Figure 5. The pictures of the flat structure, electrical field and the trajectory of the electron's movement.

Where: 1 – electrode with distributed potential (V1 V, –U2 V are the values of accelerating and braking potentials accordingly), 2 – the picture of electrical field, 3– closed trajectory of the electron. So, the opportunity of creation of oscillating system in the structure with braking electrical field, which were made by the distributed potentials and accelerating potential at the electrodes, is shown. The electron in such the field will make fourfold process of braking and accelerating. The occasion of existence of electron's cyclic rectilinear fluctuating in the suggesting electrical field is very interesting. When we divide the equations from the expression (4) one by one, we shall receive following:

$$y(t) = x(t) \qquad (26)$$

The typical trajectories are submitted at the fig. 6.

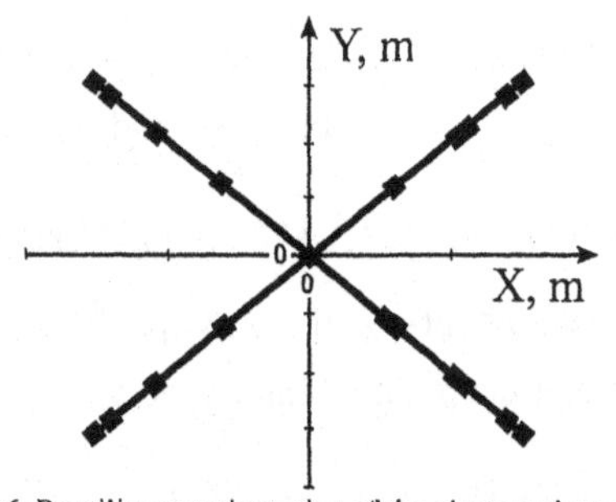

Figure 6. Rectilinear trajectories of the electrons' movement.

If we inject the electrons to the submitted structure from the angles of received square at the same time, then we shall have the opportunity of creation of generators on the oncoming beams with electromagnetic radiation's generation. If we impose transverse electromagnetic field in addition, then the electron will have to get some more cyclic spiral trajectory around closed trajectory, and this means, that the principal opportunity of creation of generator similar to the cyclic MCR-generator exists. Under creation of spherical source of electrons and creation of accelerating potential at the centre of the sphere and under synchronous injecting of electrons, we also have opportunity of generators' creation on the oncoming beams with the generating of electromagnetic radiation or we can use it for quick heating of material using instead of anode.

4. The cyclic trajectories of electron's movement in electrical field with distributed potential in the bulk structure.

For forming of apparent electrons' movement (circular helix) it is necessary to create in addition the tangential constituent of velocity Z-direction. We can create such conditions at the cost of distributed electrical field along Z-axle and then the electron in the submitted structure will move over circular helix. I.e. the using of bulk electrical field will allow creating of the generators of the electromagnetic radiation (the analogy is the laser on the free electrons) and, probably, can be and plasma trap.

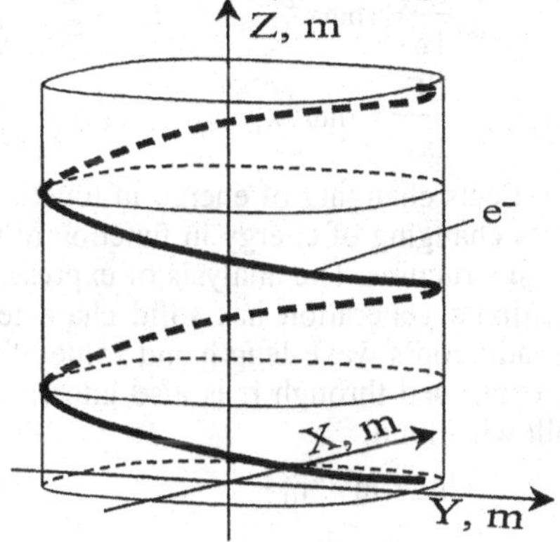

Figure 7. The trajectory of the electron's movement's in the bulk structure.

5. Concerning the generation of electromagnetic radiation.

It is known, that the process of electromagnetic energy's radiation takes place under accelerating or braking in electrical field. According to (25) we accept:

$$r_0 = x_0 \qquad (27)$$

Squaring the expression (29) and multiplying it by $m/2p^2$, we shall receive:

$$m/_2\, p^2 r^2 = m/_2\, p^2 r_0^2 \qquad (28)$$

$$m/_2\, p^2 r^2 = E \qquad (29)$$

As the physical meaning of p-parameter is the ω-frequency, then

$$J\, \omega^2/_2 = J_0\, \omega^2/_2 \qquad (30)$$

where: E - the value of kinetic energy, J_0 - the value of moment of inertia of the electron. The expression (30) coincide with the classical expressions, describing the electron's movement in the atom [4]. The process is of interest when the velocity of the electron exceeds the velocity determining by the expression (11). In this case the energy of falling particle exceeds the value of energy of full absorption by the system, i.e. the electron goes out of the limits of the structure. According to the Guygents-Shteyner's theorem [4] the full energy adds up the energy of central axial rotation and augmented A-axial rotation and writes as:

$$J\, \omega^2/_2 - J_0\, \omega^2/_2 = \pm\, ms^2\, \omega^2/_2 \qquad (31)$$

$$E_{FUL} - E_{ADS} = E_{GOOUT} \qquad (32)$$

Where: $E_{FUL}, E_{ADS}, E_{GO OUT}$ – the values of full energy, absorbed energy and energy, which the electron leaves the structure with, accordingly. So, if we know the energy of falling electron and the energy of absorption, we can always determine residual energy of the electron leaving the structure, and we can also determine its movement trajectory, and by changing of electron's movement trajectory (the result of interaction) we can tell about characteristics of the structure. Let us investigate the expression (30) for subject of energetic and spectral distribution of radiation depending on p-parameter and initial point of electron's entry:

$$\frac{dE}{d\omega} = m\omega x_0^2 \tag{33}$$

$$\frac{dE}{dx_0} = m\omega^2 x_0 \tag{34}$$

The expression (33) reflects changing of energy in function of the ω-parameter, and the expression (34) reflects changing of energy in function of the value of initial entry point x_0 of the electron to the structure. The analysis of expressions (33) and (34) shows, that the spectrum of radiation's generation has solid character. The knowledge of perfect relationship between radiation's wave-length and value of electron's entry point's to the structure coordinate, expressed through r, is also interest of science. Let us rewrite the expression (32) in following kind:

$$h\nu = m\omega^2 r_0^2 \tag{35}$$

$$\lambda = \frac{ch}{m\omega\ r_0^2} \tag{36}$$

Expression (36) of perfect relationship between radiation's wave-length and value of electron's entry point's to the structure coordinate, expressed through r. As an example we shall consider a possibility of origin of an optical radiation in an ionised medium. The optical radiation in ionised medium, derivable under passing through it charged particles, is of great scientific and practical interest. As is well known [5] the ionised medium is characterised by the presence of free electrons and ions. The braking electrical field with distributed potential in this medium can be created either by the electron, or by the group of electrons, or by the dipole, or by the system electron-ion- electron. Let's define parameters of a system with a distributed potential, in which the generation of electromagnetic radiation of an optical range is possible. For possible generation of electromagnetic radiation in an optical range in with a wavelength $\lambda \approx 5\cdot10^{-7}$ and frequency $\omega \approx 3{,}77\cdot10^{15}$ rad./ sec and for energies of a flying electron $E \approx 3{,}97\cdot10^{-19}$ joule =2,5 ev,, the sizes of a structure (proceeding from expressions (11) or (12)) should have values: $r = d \approx 1{,}7\cdot10^{-10}$ m. The braking potential about a value 2,5 V can be created by a charge of an electron, if it to consider dot, on a distance $\approx 2{,}52\cdot10^{-10}$ m. And then it is possible to assume, that the electron creates a braking field with a distributed potential with values of sizes of a structure $r = d \approx 2{,}\ 52\cdot10^{-10}$ m and in it there can be a transformation of an initial kinetic energy of an electron with a value $E \approx 2{,}5$ ev in electromagnetic radiation with $\lambda \approx 5\cdot10^{-7}$. It means, that in solid mediums, at passing of charged particles through substance circumscribed above the mechanism of origin of an optical radiation is possible. It is possible to mark, that the mechanism of appearance of optical radiation (the optical band) of hot substances also keep within limits of suggesting model, as for the electrons with heat velocities the braking field electrical field with distributed potential will always be found, for example it could forms because of micro projections (surface roughness).

6. Concerning the driving a particle in a potential field with a distributed potential (experience A. A. Michelson).

Let's consider driving a photon, as driving of a particle in a potential field with a distributed potential. We shall consider the process of driving of a photon in a field, which develops of a sum of two gravitational fields. The first field are formed by a system « the Sun - Earth », and second (considering that the mass of the Earth is concentrated in the kernel) system « the kernel of the Earth - surface of the Earth ». Proceeding from a principle of superposition we shall consider driving a photon in these fields separately. Then the plane-parallel structures with a distributed potential for these fields will have the following parameters (see. A Fig. 1). For a system « the Sun - Earth ». On intersection of a plane y = d and axes Y we shall locate the Sun and we shall accept it for a dot radiant. Distance between planes d (plane y = 0) we shall accept for a distance from the Sun up to the Earth = $150 \cdot 10^9$ m. Length of a structure (length is flat of a parallel structure L) we shall accept also for a distance from the Sun up to the Earth = $150 \cdot 10^9$ m. Value of a potential of the Earth, because of its small value on (the small contribution to a value of a parameter ð) on a plane y = 0 is accepted for 0. For a system « the kernel of the Earth - surface of the Earth ». On intersection of a plane y = d and axes Y we shall locate the kernel of the Earth also we shall accept it for a point source. A distance between planes d and length of a structure L we shall accept behind a radius of the Earth d = L =$6{,}4 \cdot 10^6$ m. At the definition of a parameter p (for a system « the kernel of the Earth - surface of the Earth ») we neglect sizes of a structure, in which the experience (distance between reflecting mirrors was carried out and interferometer made 10ì.) because of it of a small, concerning a radius of the Earth. The potential of a fields created by a skew field by a mass Ì is determined by expression:

$$\varphi = \gamma \frac{M}{r} \tag{37}$$

And accordingly equations driving of a particle are noted [5]:

$$\dot{v} = -\operatorname{grad}\varphi \tag{38}$$

Where:γ,φ, M, r - accordingly values of a gravitational constant, gravitational potential, mass of a gravitating skew field and distance between skew fields. The set of equations for the definition of trajectory of driving of a particle in a gravitational field with a distributed potential in the correspondence with [2] will look like:

$$\begin{cases} \ddot{x} = F_x = \dfrac{GM_S}{d^2} y = \alpha y \\ \ddot{y} = F_y = \dfrac{GM_S}{d^2} y = \alpha x \end{cases} \tag{39}$$

And then driving of a particle in fields with a distributed potential will be described by expressions (5), (6), (7) and (8), and the parameter p of a structure « the Sun - Earth » will be defined:

$$p_S = \sqrt{\alpha} = \sqrt{\frac{6{,}67 \cdot 10^{-11} 1{,}97 \cdot 10^{30}}{225 \cdot 10^{20}}} = 7{,}65 \cdot 10^{-2} \tag{40}$$

The parameter ð For a system the kernel of « the kernel of the Earth - surface of the Earth ».will be defined:

$$p_E = \sqrt{\alpha} = \sqrt{\frac{6{,}67 \cdot 10^{-11} 6 \cdot 10^{24}}{(6{,}4 \cdot 10^6)^2}} = \sqrt{9{,}77} \tag{41}$$

α- in the given expression, gains a physical sense of gravity.
Inasmuch as parameter of parallel-sided structure« the Sun - Earth » differs from parameter of structure, where experiment took place, for 2 degrees, we won't consider M, because of it's small influence. Path-length difference in experiment is able to give:

$$\Delta V(t) = C(\sqrt{1\text{-}2pt + 2p^2t^2} - \sqrt{1\text{-}2pt - 2p^2t^2} \approx C \cdot 10^{-10}\,\text{m/sec} \tag{42}$$

Residual of distances, which photon passes will make:

$$\Delta V \cdot t \approx C \cdot 10^{-10} \cdot 6{,}67 \cdot 10^{-8} \approx 2 \cdot 10^{-8}\,\text{m} \tag{43}$$

It means, that the residual of a course of rays can make magnitude approximately about 2 degrees less, than wavelength of a photon. It is natural, that at such value of a difference of a course of rays there is no physical possibility to receive an interference picture in carried out experience. It is necessary to expect, that the obtained value of a difference of a course in experience, probably, is maximum, as in an evaluation was not taken into account a residual of a course of rays in the most mirror slice. Possibly, acquired result is maximal, inasmuch as waiting, that allowing for results of process of reflection of light in mirror plate itself even greater deceleration take place. It is necessary to mark that fact, that the velocity of driving of ground, rather light has a small and contribution it in a residual of velocities also will be insignificant.

7. The Conclusion

The analytical regularities of electronic flow in an inhomogeneous electrical field with a distributed potential are submitted in a paper, can be used for the analysis of dynamics of the electron's driving in electrical fields and for their calculation. The represented outcomes of researches can be useful at creation of generators with cyclical circular (laser on free electrons) and cyclical rectilinear trajectories (counter flows) electromagnetic radiation in an offered symmetrical electrical field. The offered exposition of the physical process of origin in structures with a distributed potential of sites with a negative resistance (« tunnel effect »), probably, will allow to create electronic - dynamic structures with stabile parameters. In potential fields with a distributed potential does not exist uniform and rectilinear driving of particles. The relative velocity of driving of a particle in such fields depends on a value of a parameter of interaction of a field with a particle p, magnitude of an initial velocity of an entry of a particle in this field and magnitudes of an angle of the entry. The offered method of the analysis of particle's driving in potential fields, as in fields with a distributed potential, probably, allows on new to consider processes of particle's driving in such fields and to explain a residual of a course of light in experiences A. A. Michelson. Under the idea about gravitational field, in which A. A. Michelson 's experiment had place, like about field with distributed potential, theoretical analysis of his results is carried out. It is shown, difference of velocities in longitudinal and transversal directions may give $3 \cdot 10^{-7}$ size of order of value of light speed, and path-length difference may give $2 \cdot 10^{-8}$ m, i.e. value about two degree less than length of photon wavelength. Inasmuch as path-length difference is differed for value about 10^{-2} part of wavelength, receiving of fringe pattern for this case is difficult.

References:

1. Landau, L., Lifchic, E (1988)The theory of a field, Science, Moscow
2. V. I. Fedulov, A. B. Smirnon (2000) The method for estimating and transformation of electron's flow in the plane-parallel vacuum structures, Uzbek journal of physics, **5-6** , p.p. 442-446
3. V. I. Fedulov, V. I. Suvorov, U. Umirov. (2001) Concerning the trajectory of electron's flow in braking electrical fields with distributed potential, Uzbek journal of physics, **5-6** , p.p. 442-446
4. V. I. Fedulov, V. I. Suvorov, U. Umirov (2002) Concerning the radiation of electron's flow in braking electrical fields with distributed potential, Uzbek journal of physics, **3-4** , p.p. 242-246
5. H. Kuchling, (1980), Physik, VEB Fachbuchverlag, Leipzig,
6. J. V. Jelley (1958) Gerenkov radiation and application, Pergamon press, London, New York, Paris, Los Angeles
7. Lorents G.A. (1935ã) ' Electromagnetic phenomena in system moving with ane velocity smaller than of moving of light ' in ' A Principle of a relativity theory ', ÎÍÒÈ p.p.16-18

HIGH CURRENT PLASMA LENS (STATUS AND NEW DEVELOPMENTS)

A.A. Goncharov
Institute of Physics NASU
Kiev, 03039
Ukraine.

Abstract. Electrostatic Plasma Lenses may be beyond comparison as a tool for focusing and manipulating of high-current heavy ion beams. The reason for this is a principal quasi-neutrality of such beams; under these conditions it is not possible to use traditional vacuum electrostatic lenses. At the same time, employment of vacuum magneto-static lenses for heavy ions focusing requires highly expensive efforts for creation of considerable magnetic fields. Thereupon, employment of Plasma Lenses which do not require considerable electric and magnetic fields, especially for moderate energy (10-100keV) heavy ion beams focusing, may represent an attractive and unique alternative. Here we briefly review the plasma lens (PL) fundamentals and summarize some recent developments (experiments, theory, computer simulations) that has been performed at the Institute of Physics National Academy of Science of Ukraine and at the Lawrence Berkeley National Laboratory, USA. We show that there is a very narrow range of low magnetic field for which the optical properties of the lens improve drastically. This open up attractive possibilities for elaboration new generation compact PL based on permanent magnets. The first experimental results obtained in 2001 at Kiev and Berkeley on the operation of a permanent magnet PL are presented and summarized based on computer modeling stationary processes at volume PL and theoretical calculations.

1. Introduction

The high current electrostatic PL is an axially symmetric plasma-optics device with a set of cylindrical ring electrodes disposed within the magnetic field region, with field lines connecting ring electrode pairs symmetrically about the lens mid-plane. The fundamental concept of this kind of lens based on the use of magnetically insulated electrons and equipotentialization of magnetic field lines [1-2]. It implies strong magnetization of electrons under conditions $\rho_e \ll R$, where ρ_e is the electron Larmor radius and R is typical dimension of the PL. Under these conditions electrons are tightly bound to the magnetic field lines, moving freely along them. It limits the transverse

E. Oks and I. Brown (eds.),
Emerging Applications of Vacuum-Arc-Produced Plasma, Ion and Electron Beams, 227–233.

mobility of electrons and enables an introducing into the plasma volume of dc electric fields suitable for manipulation of high current beams of non-magnetized ions. The basic peculiarities of ion beam focusing by this kind device have been demonstrated in a number of experiments [3-6]. It was shown that creation of the optimum magnetic field shape, and careful selection of the external potential distribution applied to the lens electrodes corresponding to plasma optical principles give the possibility for the radial electric profile in the lens volume to be varied in wide ranges. Some features of the focusing of wide-aperture, high-current hydrogen ion beams are described in [3-4]. In the high current PL one can vary the radial profile of the electric potential in the PL volume, minimize the spherical aberrations, and thus create the required beam profile on the target, in particular, to transform an inhomogeneous ion beam profile into a homogeneous profile. It was noted that PL operation in the high current regime occurs when the beam potential parameter $I_b/4\pi\varepsilon_0 V_b$ is greater than the maximum externally applied PL voltage, $I_b/4\pi\varepsilon_0 V_b >> \varphi_L$ [5]. Here I_b is the ion beam current, V_b beam velocity and ε_0 is permitivity of free space.

In these first experiments, repetitively pulsed, broad (ϕ=5.6cm), low divergence, low noise beams of hydrogen ions were used, with total current up to 2A, energy up to 25keV, and pulse duration about 100μs We found good qualitative accordance with theory. Reliable operation of the high current PL was demonstrated, with good predictability of the device. At the same time we noted that the maximum beam compression (ratio of focused to unfocused beam current density at the focus) was restricted to disappointingly low values of 2 - 5. It was recognized some time later [6] that this restriction was connected to nonremovable momentum aberrations due to the azimuthal rotation of fast beam particles in the lens magnetic field. These aberrations depend on the ion charge-to-mass ratio through the beam velocity V_b and limit the minimum radius of the focused beam to

$$R_{\min} = R_0 \cdot \frac{v_b B L}{\varphi_b \pi c} \tag{1}$$

Here R_0 is the initial radius of the beam, c is the velocity of light, B is the magnetic field within the lens volume, L is the length of lens. One can show that beams of protons with energy 10 – 25 keV and low-energy singly-charged copper ions with energy 200 – 400 eV, for typical lens parameters (B ~ 0,1T) , will both come to a focus of radius R_{min}~1 cm. For higher energy heavy ions the effect of momentum aberrations is much reduced and, for example, for 20-30 keV copper ions a focal spot with $R_{min} \approx 1$ mm could be produced. Thus this kind of electrostatic PL can efficiently focus large area, energetic heavy ion beams.

This was demonstrated by a series of experiments performed at the Kiev Institute of Physics (IP NASU) and at Berkeley (LBNL). In these experiments we used facilities described in detail elsewhere (Kiev [7] and Berkeley [8-9]). We used a MEVVA-type vacuum arc ion source with a two-chamber anode and three-electrode multi-aperture extraction system. The sources operated in a repetitively-pulsed mode and produced moderate energy, heavy metal ion beams with parameters: Kiev - beam duration 100 μs, total current up to 800 mA, beam extraction voltage up to 25 kV, initial diameter 5.6 cm, ion species Cu, Mo, and C. Berkeley - pulse duration 250 μs, extraction voltage up to 50 kV, beam current up to 500 mA, diameter 6 or 10 cm, beam species C, Mg, Nb,

Cu, Zn, Ta, Pb, and Bi. The ion sources were located about 30 cm from the mid-plane of the PL. For our initial work we used a lens in which the magnetic field was provided by conventional current-driven coils surrounding the lens electrodes. The pulsed magnetic field strength was up to 0.1 T and pulse length was up to 500μs. Our first experiments were performed at Kiev [7]. The maximum current density of copper ions onto a collector at the focus was up to 170 mA/cm^2, and the maximum compression (of current density) for optimal conditions of minimized spherical aberrations is 20x.

Further investigations were performed at Berkeley [8-9]. The experiments explored operation of the electrostatic PL with moderate energy (10–100) keV, high current (up to amperes), large-area, heavy ion beams. It was shown that the PL can be used as a strong focusing device for creation of a spot with high energy density at the focus. For the optimal lens potential distribution the tantalum ion beam compression at the focus was a factor of 30x. We showed also that variation of the lens potential distribution can change the lens operating regime from focusing to defocusing. This defocusing regime occurs for the case of very long potential distribution, and could find application for tailoring of beam radial current density profiles.

It was considered some possible factors limiting the maximum compression of the focused ion beam. Momentum aberrations, finite phase volume of the ion beam, and incomplete space charge compensation of the beam at the focus were considered; these factors can occur for beam compressions greater than 1000 under these conditions. We concluded that the maximum beam compression observed experimentally is restricted by nonremovable spherical aberrations because of the finite width of the lens electrodes, even for optimal potential distribution. We showed that for a narrow range of low magnetic field, for which the condition $\rho_e \leq R$ is satisfied the lens properties improves drastically. For these conditions the turbulence level in the lens volume and in the focused beam decrease by more than an order of magnitude [11]. Under this the radial potential distribution at the PL mid-plane is close to parabolic, for which spherical aberrations are eliminated. At the same time, the maximum beam compression at the focus significantly exceeds the values obtained for typical PL magnetic field strengths.

. This preliminary work provided the background for the development of a new generation compact PL based on using of permanent magnets rather than an electrically driven coils surrounding the lens region. Here we describe some new results, including experimental investigations, theoretical analyses and computer simulation of the formation of the plasma in this kind of PL.

2. EXPERIMENTAL CONDITIONS AND APPROACH

The experiments were carried out collaboratively at Kiev and at Berkeley.

The basic parameters of the Plasma lenses used at the experiments were as follows. Kiev – input aperture D = 7.4 cm, length L = 14 cm, number of electrostatic electrodes N = 13; the electrodes were fed via an RC-divider that provided fixed electrode potentials for the duration of the ion beam, and the highest potential (U_L) applied to the central electrode was + 4.7 kV; the maximum strength of the magnetic field formed by the permanent Fe-Nd-B magnets at the center of the lens was B = 360 G. Berkeley – D = 10 cm, L = 15 cm, N = 11; the electrodes were fed by a 110 kΩ resistive voltage divider; $U_L \leq +10$ kV, B = 300 G. The magnetic field shape required for each PL and the corresponding disposition of magnets need to establish the magnetic field were determined by computer simulation and experimental tests. The magnetic field strength B could be varied by changing the number of magnets used and also by employing iron pieces to shunt a part of the magnetic field, allowing the field strength to be changed by increments of 17 G.

Radially movable Langmuir probes were used for measurement of the plasma in the lens volume and in the beam drift space. I_b and J_b were measured by an axially-movable sectioned collector (at Kiev) and by a radially-movable, magnetically-suppressed Faraday cup with entrance aperture 3 mm (at Berkeley), located at a distance ~30 cm from the lens mid-plane. The base pressure in the vacuum chamber was less than 1×10^{-5} Torr, allowing formation of plasma within the PL volume by the ion beam itself and by secondary electron emission electrons from the lens electrodes.

3. RESULTS

It was investigated the static and dynamic characteristics and focusing properties of this kind of PL as a function of magnetic field, externally applied electrode potential distribution, total transported ion beam current, initial beam diameter and ion species. It was found that this optimal focusing regime is sensitive to magnetic field and depends strongly on the plasma formation conditions in the PL volume and on the ratio of initial beam diameter to PL input aperture. The experiments show that for a large area ion beam with initial beam diameter equal to the input aperture of the PL, the focusing properties are more distinct. The maximum compression for the tantalum ion beam was a factor of 5-7 for the optimal lens potential distribution, for the case of beam with initial diameter 6 cm. At the same time for the case of tantalum beam with diameter 10 cm, the maximum beam compression at the PL focus was approximately a factor of 20, with and current density up to 32 mA/cm^2. Note that similar results were obtained for a copper ion beam on the Kiev set-up, where the compression was a factor 15-25 depending on the total ion beam current passing through the lens (see [10] for details). Note also that similar results were observed when beams of this type were focussed using a PL with conventional current-driven coils [9].

The experimental results depend on the particular externally-applied potential distribution along the lens electrodes. The optimal distribution minimizes lens spherical aberrations, as established empirically for a PL with an input aperture of 10 cm., and this distribution differs significantly from the theoretical optimum distribution obtained by plasma optic principles.

Focusing of different ion beams species (Bi, Pb, Ta, Nb, Mg, Cu, C) was investigated. Better results were obtained for the case of the heavy ion beams Bi, Pb, and Ta. In Fig. 1 one can see good bismuth beam compression for the case of the optimal distribution. Fig.2 show that the radial profile of the focused Bi ion beam depends on the PL operation, for case of electrode optimal potential distribution. Note that the maximum ion beam compression for Bi was up to a factor of 30, and the low-noise focused beam current density was up to 45mA/cm^2.

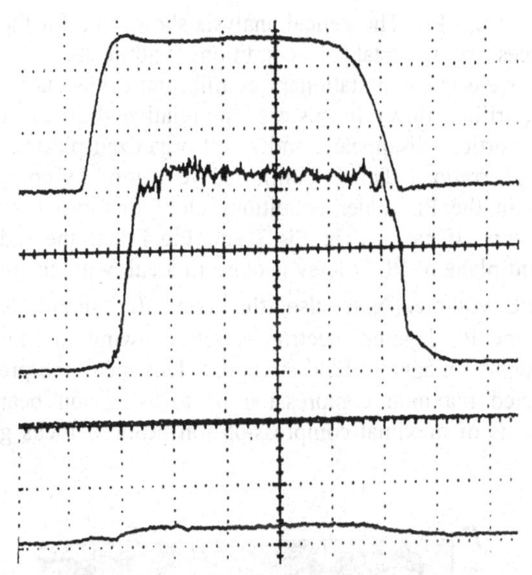

Figure1. Oscillograms of bismuth beam current measured by an on-axis Faraday cup, for the case of maximum compression and the experimental optimal PL potential distribution. Beam accelerating voltage U_{acc} = 34kV, U_L=7.7kV. B=300G. Upper trace – ion source arc current (200A/cm); middle trace – ion beam current density for cases PL on; lower trace - PL-off. Vertical scale: 14mA/cm^2 per cm. Horizontal scale: 50μs/cm.

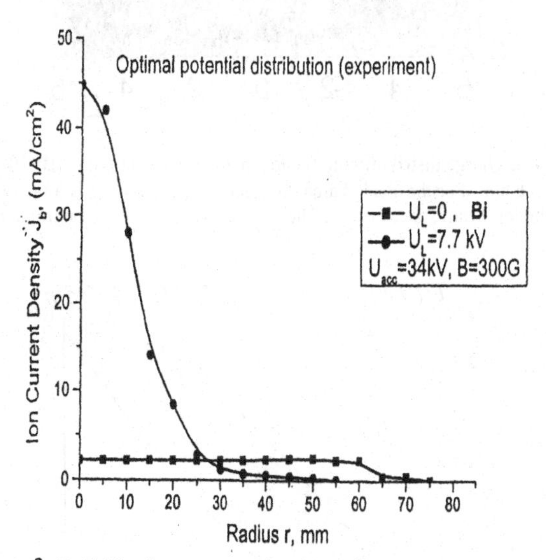

Figure 2. Radial ion beam current density profile at the Faraday cup location.

These results and experimental conditions were used for theoretical analysis and computer simulation of the processes of formation of the plasma medium within the electrostatic high current PL. Theoretical analysis shows that for the optimal operating regime, it is necessary to establish conditions within the PL volume that will overcompensate for electrons in stationary equilibrium orbits such as axial symmetric equilibrium flow (Brillion flow). In this case the relative sliding of electron layers that leads to turbulent vortices disappears, and a self-organized plasma medium is formed without spherical aberrations. This is confirmed very good by computer simulation of plasma formation in the PL under conditions close to those used in the Kiev and Berkeley experiments. It can see in Fig.3 and Fig.4 that the radial distribution of electrons in the mid-plane of PL at Kiev is close to ideal, without spherical aberrations, in most of the PL volume. Note also, that computer simulation demonstrate the formation in volume PL layered electron structure owing to finite width ring lens electrodes under typical magnetic field strengths. This mean the presence of spherical aberrations restricted maximal compression of focusing ion beam. The results of computer simulations of maximal compression ion beam at focus good coincide with experimental data.

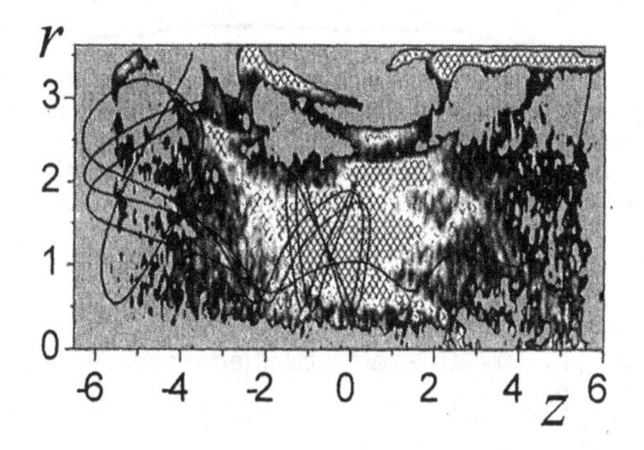

Figure 3. Electron space charge distribution in the low magnetic field regime (B=100G, U_{acc}=12kV, U_L= 3kV, PL input aperture 3.6 cm). For the shaded area the electron space charge is a factor 1.5 or more greater than the ion beam space charge. The solid curve is one electron trajectory

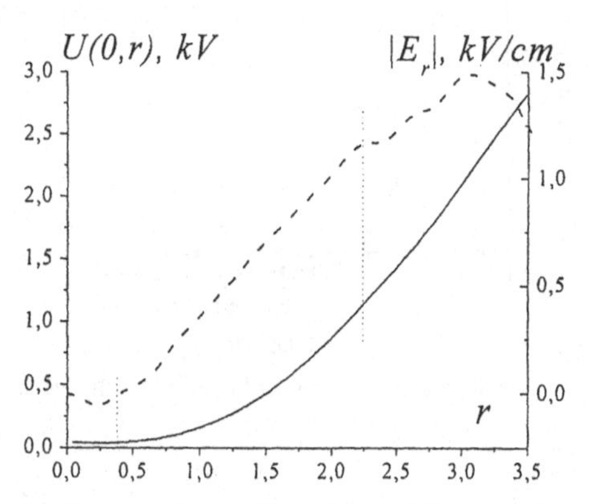

Figure 4. Radial potential distribution (solid curve) and electric field (dotted line) in the PL mid-plane.

In summary, here we have presented results the novel high current Plasma Lens development. A simple design, the robust construction, only one power supply and the high efficiency are attractive advantages of the PL based on the use of permanent magnets rather than conventional current-driven coils. For this stage this device can be used, for example, in particle accelerator beam lines and in high dose ion implantation facility. In the meantime it need further research complex efforts (experimental, theoretical and computer simulations) on creation optimal PL without spherical aberrations, in part, by optimisation magnetic field line configuration in the range of low magnetic fields. This could open unexpected and attractive areas for the application of high current heavy ion beams.

Acknowledgements

The author is thankful for the great contribution to this work from Ivan Protsenko, Vaycheslav Gorshkov, Sergey Gubarev, Andrey Dobrovolskiy, Irina Litovko and Georgiy Yushkov at various stages of these investigations and also grateful very much to Dr. Ian Brown for beneficial discussions and fruitful collaboration for a all last years.. We wish to acknowledge the financial support from the Science and Technology Centre in Ukraine through project No.1596.

References:

1. A. Morozov, *Dokl. Acad. Nauk USSR,* **163,** 1363 (1965).
2. A. Morozov and S. Lebedev, *Reviews of Plasma Physics*, M. Leontovich, Ed., New York: Consultants Bureau, 247 (1975).
3. A.Goncharov, A.Dobrovolsky, A.Zatuagan and I.Protsenko, *IEEE Trans. Plasma Sci*, **21**, 573 (1993).
4. A.Goncharov, A.Zatuagan and I.Protsenko, *IEEE Trans. Plasma Sci.*, **21**, 578 (1993).
5. A. Goncharov, *Rev. Sci. Instrum.*, **69**, 1150 (1998).
6. A.Goncharov, A.Dobrovolsky, I. Litovko, I.Protsenko and V.Zadorodzny, *IEEE Trans. Plasma Sci*, **25**, 709 (1997).
7. A.Goncharov, S.Gubarev, A.Dobrovolsky, I. Protsenko, I..Litovko and I.Brown, *IEEE Trans. Plasma Sci*, **27**, 1068 (1999).
8. A.Goncharov, I.Protsenko, G. Yushkov and I.Brown, *Appl.Phys. Lett.*, 75, 911 (1999).
9. A. Goncharov, I.Protsenko , G. Yushkov and I. Brown, *IEEE Trans. Plasma Sci*, **28**, 2238 (2000).
10. A.Goncharov, V.Gorshkov, S.Gubarev, A.Dobrovosky, I.Protsenko and I.Brown, *Rev. Sci. Instrum.*, **73**, 1001 (2002).
11. A,Goncharov, *Rev. Sci. Instrum.*, **73**, 1004 (2002).

INDEX

Printed in the United States
By Bookmasters

Printed in the United States
By Bookmasters